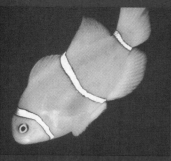

STAMS® SERIES
C

STRATEGIES TO ACHIEVE MATHEMATICS SUCCESS

- ☐ PROVIDES INSTRUCTIONAL ACTIVITIES FOR 12 MATHEMATICS STRATEGIES
- ☐ USES A STEP-BY-STEP APPROACH TO ACHIEVE MATHEMATICS SUCCESS
- ☐ PREPARES STUDENTS FOR SELF-ASSESSMENT IN MATHEMATICS COMPREHENSION

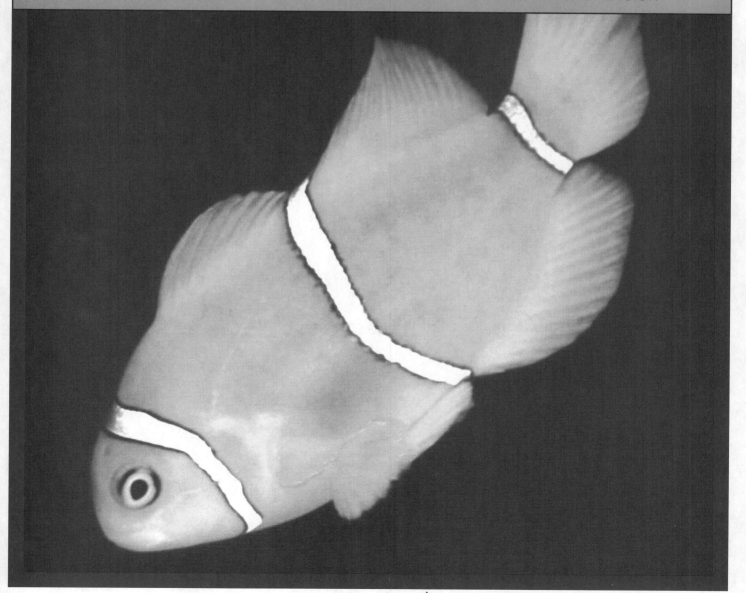

CURRICULUM ASSOCIATES®, INC.

Acknowledgments

Product Development by Chameleon Publishing Services.
Revision (2006) by Robert G. Forest, EdD.

ISBN 978-0-7609-3649-8
©2006, 2001—Curriculum Associates, Inc.
North Billerica, MA 01862
No part of this book may be reproduced by any means
without written permission from the publisher.
All Rights Reserved. Printed in USA.
15 14 13 12 11 10 9 8 7 6 5 4

TABLE OF CONTENTS

Strategy **One**	Building Number Sense	4
Strategy **Two**	Using Estimation	14
Strategy **Three**	Applying Addition	24
Strategies **One–Three**	REVIEW	34
Strategy **Four**	Applying Subtraction	38
Strategy **Five**	Applying Multiplication	48
Strategy **Six**	Applying Division	58
Strategies **Four–Six**	REVIEW	68
Strategy **Seven**	Converting Time and Money	72
Strategy **Eight**	Converting Customary and Metric Measures	82
Strategy **Nine**	Using Algebra	92
Strategies **Seven–Nine**	REVIEW	102
Strategy **Ten**	Using Geometry	106
Strategy **Eleven**	Determining Probability and Averages	116
Strategy **Twelve**	Interpreting Graphs and Charts	126
Strategies **Ten–Twelve**	REVIEW	136
Strategies **One–Twelve**	FINAL REVIEW	140

Strategy One: BUILDING NUMBER SENSE

PART ONE: Think About Number Sense

WHAT DO YOU KNOW ABOUT NUMBER SENSE?

A number is a basic part of math.

A number has a value.

Numbers can be written in numerals, called *digits*, like 6 and 9.

Numbers can be written in words, like "six" and "nine."

▶ What are the ten digits used in math?

The ten digits are 0, 1, 2, ____, ____, ____, ____, ____, ____, ____.

Each digit in a number has *place value*. The value of a digit depends on its place in a number. The number 23 is made from two digits. Look at the position of each digit. The digit 2 is in the *tens* place and has a value of 2 tens or 20. The digit 3 is in the *ones* place and has a value of 3 ones or 3.

$$23 = 2 \text{ tens} + 3 \text{ ones}$$

▶ Write three things you know about the number 34.

a. _____

b. _____

c. _____

▶ Write three things you know about the number 234.

a. _____

b. _____

c. _____

You just wrote about numbers and place value.

WHAT DO YOU KNOW ABOUT PLACE VALUE?

Look at the number 5,342.

The digit 5 is in the *thousands* place.   The digit 5 has a value of 5,000.

The digit 3 is in the *hundreds* place.   The digit 3 has a value of 300.

The digit 4 is in the *tens* place.   The digit 4 has a value of 40.

The digit 2 is in the *ones* place.   The digit 2 has a value of 2.

> 5,342 = 5,000 + 300 + 40 + 2

▶ What is the value of each digit in the number 514?

_____
_____
_____

▶ What is the value of each digit in the number 6,173?

_____
_____
_____
_____

▶ Write the number 5,342 in words.

_____
_____

> You just reviewed place value and writing numbers.

Together, write four familiar numbers. They could include the number of your house, the number of students in your class, or the number of pages in your math book. Take turns telling at least three things you know about each number.

Building Number Sense

PART TWO: Learn About Number Sense

Study the place-value chart. It shows the number of books that students in Anna's school read during summer vacation. As you study, think about the place value of each digit. Also think about whether the number is even or odd.

thousands (1,000)	hundreds (100)	tens (10)	ones (1)
2,	5	4	9

The number 2,549 is written as two thousand, five hundred forty-nine.
The number 2,549 has four digits.
The number 2,549 has 2 thousands, 5 hundreds, 4 tens, and 9 ones.

If a number has a 0, 2, 4, 6, or 8 in the ones place, the number is even.
The number 3,792 is even.
If a number has a 1, 3, 5, 7, or 9 in the ones place, the number is odd.
The number 2,549 is odd.

You use **number sense** when you think about the place value of each digit in a number. You also use number sense to recognize whether a number is even or odd.

▶ Each digit in a number has a place value, such as ones, tens, hundreds, or thousands. The value of a digit depends on its place in a number.

▶ Even numbers have the digit 0, 2, 4, 6, or 8 in the ones place.
Odd numbers have the digit 1, 3, 5, 7, or 9 in the ones place.

▶ A number can be written in digits or in words.

Anna made a place-value chart to show how many people live in her town. Study Anna's chart. Think about the place value of each digit in the number on the chart. Also think about whether the number is even or odd. Then do Numbers 1 through 4.

thousands (1,000)	hundreds (100)	tens (10)	ones (1)
3,	4	6	8

1. What number does the chart show?
   - Ⓐ 34
   - Ⓑ 368
   - Ⓒ 3,468
   - Ⓓ 13,468

2. What is the value of the 4 in the number 3,468?
   - Ⓐ 4
   - Ⓑ 40
   - Ⓒ 400
   - Ⓓ 4,000

3. Which of these is the written form of the number 3,468?
   - Ⓐ three thousand, four hundred sixty-eight
   - Ⓑ three hundred eighty-six
   - Ⓒ thirty-four thousand, sixty-eight
   - Ⓓ eight thousand, six hundred forty-three

4. The number 3,468 is an even number. Which of these numbers is also even?
   - Ⓐ 2,006
   - Ⓑ 151
   - Ⓒ 5,309
   - Ⓓ 87

Talk about your answers to questions 1–4. Tell why you chose the answers you did.

Building Number Sense

PART THREE: Check Your Understanding

Remember: You use number sense when you think about the place value of each digit in a number. You also use number sense to determine whether a number is even or odd.

▶ Each digit in a number has a place value, such as ones, tens, hundreds, or thousands. The value of a digit depends on its place in a number.

▶ Even numbers have the digit 0, 2, 4, 6, or 8 in the ones place. Odd numbers have the digit 1, 3, 5, 7, or 9 in the ones place.

▶ A number can be written in digits or in words.

Solve this problem. As you work, ask yourself, "What does the place of each digit in a number tell me about its value?"

5. The library in Anna's school gave out 1,796 new library cards last year. What is the value of the 7 in 1,796?
   - Ⓐ 7
   - Ⓑ 70
   - Ⓒ 700
   - Ⓓ 7,000

Solve another problem. As you work, ask yourself, "What number has a 0, 2, 4, 6, or 8 in the ones place? What number has a 1, 3, 5, 7, or 9 in the ones place?"

6. Anna wrote four groups of numbers. Which of the groups contains only one even number?
   - Ⓐ 19, 21, 28, 6
   - Ⓑ 14, 5, 23, 17
   - Ⓒ 15, 2, 16, 30
   - Ⓓ 31, 29, 12, 4

Building Number Sense

**Look at the answer choices for each question.
Read why each answer choice is correct or not correct.**

5. The library in Anna's school gave out 1,796 new library cards last year. What is the value of the 7 in 1,796?

   Ⓐ 7

   This answer is not correct because 7 is equal to 7 ones. The 7 is not in the ones place in 1,796.

   Ⓑ 70

   This answer is not correct because 70 is equal to 7 tens. The 7 is not in the tens place in 1,796.

   ● 700

   This answer is correct because 700 is equal to 7 hundreds. The 7 is in the hundreds place in 1,796.

   Ⓓ 7,000

   This answer is not correct because 7,000 is equal to 7 thousands. The 7 is not in the thousands place in 1,796.

6. Anna wrote four groups of numbers. Which of the groups contains only one even number?

   Ⓐ 19, 21, 28, 6

   This answer is not correct because there are two even numbers: 28 and 6.

   ● 14, 5, 23, 17

   This answer is correct because it contains only one even number. The even number is 14.

   Ⓒ 15, 2, 16, 30

   This answer is not correct because there are three even numbers: 2, 16, and 30.

   Ⓓ 31, 29, 12, 4

   This answer is not correct because there are two even numbers: 12 and 4.

PART FOUR: Learn More About Number Sense

You use number sense when you count.

▶ By counting, you know what number comes before or after another number.

▶ To tell the place in order of an item in a row, line, list, or other group, use the ordinal numbers *first, second, third, fourth, fifth,* and so forth.

Anna helped Uncle Val count the stamps in his stamp collection. Do Numbers 7 through 10.

7. Anna counted 215 stamps in one of Uncle Val's stamp albums. If Anna's uncle had put 1 less stamp in the album, how many stamps would Anna have counted?
   Ⓐ 216 stamps
   Ⓑ 210 stamps
   Ⓒ 220 stamps
   Ⓓ 214 stamps

8. Anna's uncle had 6 stamps on one page. What place in line is the stamp with a star?

Start

   Ⓐ fourth
   Ⓑ sixth
   Ⓒ second
   Ⓓ fifth

9. Uncle Val gave Anna 126 stamps to start her own collection. If she had 1 more stamp, what number would she write in her record book?
   Ⓐ 125
   Ⓑ 127
   Ⓒ 130
   Ⓓ 120

10. Anna bought 7 heart stamps. In what place is the shaded-heart stamp?

Start

   Ⓐ seventh
   Ⓑ fourth
   Ⓒ fifth
   Ⓓ third

Building Number Sense

Read the chart that Anna made after she helped her teacher count supplies. Then do Numbers 11 through 14.

The supply closet has three shelves. The chart shows the number of items on each shelf.

SHELF	ITEMS	HOW MANY
Top Shelf	• pads of paper • dictionaries	57 9
Middle Shelf	• colored markers • colored pencils • paintbrushes	103 414 38
Bottom Shelf	• erasers • boxes of chalk	32 34

11. Anna counted 32 erasers and 34 boxes of chalk. What number comes after 32 and before 34?
    Ⓐ 30
    Ⓑ 33
    Ⓒ 35
    Ⓓ 36

12. Beginning with the top shelf, what item did Anna list fifth on the chart?
    Ⓐ boxes of chalk
    Ⓑ paintbrushes
    Ⓒ dictionaries
    Ⓓ erasers

13. Anna forgot to count 1 colored pencil. What number should she have written on the chart?
    Ⓐ 413   Ⓒ 415
    Ⓑ 410   Ⓓ 424

14. Anna took 7 pads of paper from the shelf. In what place is the shaded pad?

Start

Ⓐ fifth
Ⓑ seventh
Ⓒ third
Ⓓ sixth

Building Number Sense

PART FIVE: Prepare for a Test

▶ A test question about number sense may ask for the value of a digit in a number.

▶ A test question about number sense may ask whether a number is even or odd.

▶ A test question about number sense may ask what number comes before or after another number.

▶ A test question about number sense may use the ordinal numbers *first, second, third, fourth, fifth,* and so forth, to ask the place in order of something in a row or line.

Anna learned some facts about sea turtles. Read the facts. Then do Numbers 15 and 16.

Anna Learns About Sea Turtles

Anna read that a mother sea turtle lays up to 200 eggs in a nest on a sandy beach. The eggs are as small as Ping-Pong balls. In 6 to 10 weeks, the eggs hatch. The tiny newborn turtles crawl down to the sea, where they spend the rest of their life.

Building Number Sense

15. Anna read that the largest sea turtle ever found was a leatherback turtle. It weighed more than 2,000 pounds. The exact weight is an even number. Which of these could be the exact weight?

    Ⓐ 2,135 pounds
    Ⓑ 2,328 pounds
    Ⓒ 2,019 pounds
    Ⓓ 2,107 pounds

Building Number Sense

16. Anna asked people in her town to donate money to help protect sea turtles. The number of people who gave money was 1,348. What is the value of the 3 in 1,348?

    Ⓐ 3
    Ⓑ 30
    Ⓒ 300
    Ⓓ 3,000

Building Number Sense

Anna's class went on a field trip to pick apples. Read the story Anna wrote for the third-grade newsletter. Then do Numbers 17 and 18.

A Visit to an Apple Orchard

Last Tuesday, our class visited the Morales Apple Orchard. Ms. Morales told us that her family has owned the orchard for 105 years. They have more than 800 trees and grow 10 different kinds of apples. Each fall, thousands of people come to pick ripe red apples from the trees. Our class picked a total of 184 apples.

Building Number Sense

17. Anna and the other students again counted the apples they picked. This time, they counted 1 less apple. How many apples did the class pick?
    Ⓐ 190 apples
    Ⓑ 180 apples
    Ⓒ 185 apples
    Ⓓ 183 apples

Building Number Sense

18. After they picked apples, the students in Anna's class lined up their bags of apples. In what place is the shaded bag of apples?

    Ⓐ third
    Ⓑ sixth
    Ⓒ fifth
    Ⓓ seventh

Strategy Two: USING ESTIMATION

PART ONE: Think About Estimation

WHAT DO YOU KNOW ABOUT ESTIMATING TWO-DIGIT AND THREE-DIGIT NUMBERS?

An estimate is a number that is close to the number you are looking for.

An estimate is a careful guess about what an answer will be.

An estimate may be a number that is rounded to the nearest ten or the nearest hundred.

If a number is halfway between two tens or two hundreds, it is rounded to the greater ten or the greater hundred.

▶ Answer each question on the line.
   a. Is the number 18 closer to 10 or 20? _____
   b. Is the number 23 closer to 20 or 30? _____
   c. Is the number 35 closer to 30 or 40? _____

▶ There are ten two-digit numbers that can be rounded to 60. The numbers are 55, 56, ____, ____, ____, ____, ____, ____, ____, ____.

> You just rounded numbers to the nearest ten.

▶ Write four three-digit numbers that, when rounded to the nearest hundred, would be 200. _____, _____, _____, _____

▶ Answer each question on the line.
   a. Is the number 118 closer to 100 or 200? _____
   b. Is the number 173 closer to 100 or 200? _____
   c. Is the number 249 closer to 200 or 300? _____

> You just rounded numbers to the nearest hundred.

Using Estimation

WHAT DO YOU KNOW ABOUT ESTIMATING LARGER NUMBERS?

Look at the number 6,478.

Think about the following estimates:

Rounded to the nearest *ten*, the estimate is 6,480.
Rounded to the nearest *hundred*, the estimate is 6,500.
Rounded to the nearest *thousand*, the estimate is 6,000.

▶ Why is each estimate above correct? The first explanation has been completed for you.

   a. 6,480 In the number 6,478, the 78 is closer to 80 than to 70.
   b. 6,500 _____
   c. 6,000 _____

▶ What is the correct estimate for each given number? Circle each answer.

	Number	Rounded to	Estimates
a.	4,761	the nearest 10	4,760 or 4,770
b.	10,542	the nearest 100	10,500 or 10,600
c.	22,784	the nearest 1,000	22,000 or 23,000
d.	31,365	the nearest 1,000	31,000 or 32,000
e.	13,742	the nearest 100	13,700 or 13,800

> You just reviewed information about rounding numbers.

> Together, write four five-digit numbers. Then, working alone, round each number to the nearest ten, the nearest hundred, and the nearest thousand. Take turns sharing and discussing your responses.

Using Estimation

PART TWO: Learn About Estimation

Study the numbers in the chart that Carl made. As you study, think about how to find an estimate, a number that is close to an exact number.

Number	Nearest 10	Nearest 100	Nearest 1,000	Nearest 10,000
12	10			
287	290	300		
7,312	7,310	7,300	7,000	
49,194	49,190	49,200	49,000	50,000
681,839	681,840	681,800	682,000	680,000

You use **estimation** to find the nearest ten, hundred, thousand, or ten thousand of a number.

You use estimation to check if a sum of numbers makes sense. First, you round the numbers you are adding. If you round the numbers 47 and 92 to the nearest ten, you get 50 and 90. Then you add the rounded numbers. If you add 50 + 90, you get 140, which is the estimated sum of 47 + 92. The estimated sum of 140 is very close to the actual sum of 139.

You use estimation to find a number that is close to another number. You also use estimation to check if a sum of numbers makes sense.

▶ To round a number, find its nearest ten, hundred, thousand, or ten thousand.

▶ To estimate a sum, first round the numbers you are adding. Then add the rounded numbers to get the estimate of the sum.

Carl made a chart to show information about oceans. Study Carl's chart. Think about finding the nearest ten, hundred, thousand, or ten thousand of each number in the chart. Then do Numbers 1 through 4.

Ocean	Average Depth	Deepest Point	Widest Point
Pacific	13,780 feet	37,808 feet	10,999 miles
Atlantic	10,827 feet	31,365 feet	5,965 miles
Indian	12,795 feet	29,528 feet	5,965 miles

1. Carl learned that the average depth of the Pacific Ocean is 13,780 feet. What number is 13,780 rounded to the nearest ten thousand?
   - Ⓐ 13,000
   - Ⓑ 13,700
   - Ⓒ 10,000
   - Ⓓ 20,000

2. Carl noticed that the Atlantic and the Indian Oceans are both 5,965 miles at their widest points. What number is 5,965 rounded to the nearest thousand?
   - Ⓐ 5,900
   - Ⓑ 6,000
   - Ⓒ 6,100
   - Ⓓ 5,800

3. The Atlantic Ocean is 31,365 feet at its deepest point. What thousand is closest to 31,365?
   - Ⓐ 30,000
   - Ⓑ 31,000
   - Ⓒ 32,000
   - Ⓓ 33,000

4. Carl drew pictures of ocean creatures. He drew 7 creatures for each ocean in his chart. Estimate, to the nearest ten, the number of pictures he drew.
   - Ⓐ About 10 pictures
   - Ⓑ About 20 pictures
   - Ⓒ About 30 pictures
   - Ⓓ About 40 pictures

Talk about your answers to questions 1–4. Tell why you chose the answers you did.

Using Estimation

PART THREE: Check Your Understanding

Remember: You use estimation to find a number that is close to another number. You also use estimation to check if a sum of numbers makes sense.

▶ To round a number, find its nearest ten, hundred, thousand, or ten thousand.

▶ To estimate a sum, first round the numbers you are adding. Then add the rounded numbers to get the estimate of the sum.

Solve this problem. As you work, ask yourself, "To the nearest hundred, what is the rounded number of each number being added?"

5. Carl's soccer team needs new equipment. New goals will cost $727. New uniforms will cost $384. To the nearest hundred, how much money does the team need in all?
   Ⓐ $900
   Ⓑ $1,000
   Ⓒ $1,100
   Ⓓ $1,200

Solve another problem. As you work, ask yourself, "What is the nearest ten, hundred, thousand, or ten thousand of the number?"

6. The Youth Soccer Association in Carl's city sent out about 172,000 sign-up forms, one to each house in the city. What number, when rounded to the nearest thousand, is 172,000?
   Ⓐ 171,894
   Ⓑ 162,633
   Ⓒ 170,602
   Ⓓ 176,459

Using Estimation

**Look at the answer choices for each question.
Read why each answer choice is correct or not correct.**

5. Carl's soccer team needs new equipment. New goals will cost $727. New uniforms will cost $384. To the nearest hundred, how much money does the team need in all?

    Ⓐ $900

    This answer is not correct because $727 + $384, when rounded, is $700 + $400. The sum of these estimates is $1,100, not $900.

    Ⓑ $1,000

    This answer is not correct because $727 + $384, when rounded, is $700 + $400. The sum of these estimates is $1,100, not $1,000.

    ● $1,100

    This answer is correct because $727 + $384, when rounded, is $700 + $400. The sum of these estimates is $1,100.

    Ⓓ $1,200

    This answer is not correct because $727 + $384, when rounded, is $700 + $400. The sum of these estimates is $1,100, not $1,200.

6. The Youth Soccer Association in Carl's city sent out about 172,000 sign-up forms, one to each house in the city. What number, when rounded to the nearest thousand, is 172,000?

    ● 171,894

    This answer is correct because 171,894 to the nearest thousand rounds to 172,000.

    Ⓑ 162,633

    This answer is not correct because 162,633 to the nearest thousand rounds to 163,000.

    Ⓒ 170,602

    This answer is not correct because 170,602 to the nearest thousand rounds to 171,000.

    Ⓓ 176,459

    This answer is not correct because 176,459 to the nearest thousand rounds to 176,000.

Using Estimation

PART FOUR: Learn More About Estimation

You can estimate sums of money to the nearest 10¢.

▶ Look at the exact amounts of money, and determine the number of cents. For $13.82, the number of cents is 82. For $1.27, the number of cents is 27.

Round these numbers to the nearest ten. The nearest ten of 82 is 80. The nearest ten of 27 is 30.

Replace the exact number of cents with the rounded numbers. The amount $13.82, rounded to the nearest 10¢, is $13.80. The amount $1.27, rounded to the nearest 10¢, is $1.30. The estimate of the sum of $13.82 and $1.27, to the nearest 10¢, is $15.10. ($13.80+ $1.30 = $15.10)

Carl and his family had a yard sale. Do Numbers 7 through 10.

7. Carl sold two books. He sold one book for $1.89 and the other for $2.38. To the nearest 10¢, what was the total cost of the two books?
   Ⓐ $4.10
   Ⓑ $4.00
   Ⓒ $3.90
   Ⓓ $4.30

8. Carl's father sold an old lamp for $4.22. To the nearest 10¢, how much did the lamp cost?
   Ⓐ $4.20
   Ⓑ $4.30
   Ⓒ $4.40
   Ⓓ $4.00

9. Mrs. Little bought 4 boxes of baseball cards from Carl's brother. Each box cost $3.18. To the nearest 10¢, how much did she spend on the baseball cards?
   Ⓐ $12.40
   Ⓑ $12.00
   Ⓒ $12.80
   Ⓓ $12.90

10. One shopper spent about $28.50 for Carl's old desk. What amount is $28.50 rounded to the nearest dollar?
    Ⓐ $28.59
    Ⓑ $29.00
    Ⓒ $28.49
    Ⓓ $28.00

Read a part of the report that Carl wrote about the mail. Then do Numbers 11 through 14.

People have been sending mail for over 4,000 years. Mail was first used in Egypt. In 1860 and 1861, Pony Express riders, on horseback, carried mail across the American West. Later, trains carried mail across the country. In 1918, airmail was first used in America. By the 1930s, people could send mail all over the world by airmail.

The cost of airmailing a five-pound package to different countries today is shown in the chart.

Country	Airmail Rate
Brazil	$30.43
Canada	$15.84
Mexico	$17.16
Greece	$20.88

11. To the nearest 10¢, what is the airmail rate for a five-pound package to Mexico?
Ⓐ $17.00
Ⓑ $17.10
Ⓒ $17.20
Ⓓ $17.30

12. To the nearest 10¢, what is the price of airmailing one five-pound package to Canada?
Ⓐ $15.00
Ⓑ $15.90
Ⓒ $16.00
Ⓓ $15.80

13. To the nearest 10¢, how much would it cost to send a five-pound package to Brazil?
Ⓐ $30.40
Ⓑ $30.00
Ⓒ $31.00
Ⓓ $31.40

14. To the nearest 10¢, what is the cost of airmailing a five-pound package to Greece?
Ⓐ $20.00
Ⓑ $21.00
Ⓒ $20.90
Ⓓ $21.90

Using Estimation

PART FIVE: Prepare for a Test

▶ A test question about estimation may ask for the nearest ten, hundred, thousand, or ten thousand of a number.

▶ A test question about estimation may ask for an estimated sum.

▶ A test question about estimation may ask for an amount of money estimated to the nearest 10¢.

Carl learned some facts about the tallest buildings in the world. Read the facts. Then do Numbers 15 and 16.

Tall Buildings

Carl learned that the tallest building in the United States is the Sears Tower, in Chicago. The Sears Tower is 1,454 feet tall. Carl learned that the tallest building in the world is the Petronas Tower, in Malaysia. The Petronas Tower is 1,483 feet tall. Carl also found out that the tallest lighthouse in the world is located in Japan. It is 348 feet tall.

Using Estimation

15. Carl figured out that the Sears Tower is 17,448 inches tall. Which of these is 17,448 rounded to the nearest thousand inches?

    Ⓐ 17,400
    Ⓑ 17,450
    Ⓒ 17,000
    Ⓓ 7,000

Using Estimation

16. Carl estimated the sum of the heights of the three buildings. To the nearest hundred feet, what was the sum?

    Ⓐ 3,100 feet
    Ⓑ 3,200 feet
    Ⓒ 3,300 feet
    Ⓓ 3,400 feet

Carl visited a recycling center. Read what Carl learned. Then do Numbers 17 and 18.

The Recycling Center

Carl learned that the recycling center sometimes pays for trash! He found out that the center pays $1.82 for 100 pounds of old newspapers. He also learned that the center pays $6.19 for 100 pounds of aluminum cans.

Using Estimation

17. Carl's family has 100 pounds of old newspapers in their garage. To the nearest 10¢, how much money would the recycling center pay for the newspapers?
    - Ⓐ $1.00
    - Ⓑ $1.50
    - Ⓒ $1.80
    - Ⓓ $1.90

Using Estimation

18. Carl's class raised money by collecting aluminum cans. The students collected 400 pounds of cans. To the nearest 10¢, how much money did the recycling center pay for the cans?
    - Ⓐ $24.00
    - Ⓑ $24.80
    - Ⓒ $24.40
    - Ⓓ $25.00

Strategy Three: APPLYING ADDITION

PART ONE: Think About Addition

WHAT DO YOU KNOW ABOUT ADDITION?

Addition is one of the four basic operations.

The numbers to be added are called *addends*.

Addition is used to find the total of two or more addends.

The answer in addition is called the *sum*.

To do addition correctly, you must know the basic addition facts.

▶ Write the sum for each addition fact.

a. 6 + 9 = ___   c. 7 + 9 = ___   e. 5 + 8 = ___   g. 4 + 9 = ___
   9 + 6 = ___   9 + 7 = ___   8 + 5 = ___   9 + 4 = ___

b. 8 + 4 = ___   d. 3 + 9 = ___   f. 2 + 9 = ___   h. 5 + 7 = ___
   4 + 8 = ___   9 + 3 = ___   9 + 2 = ___   7 + 5 = ___

▶ What do you notice about each pair of problems above?
Changing the order of the _____ does not change the _____.

▶ Rewrite each addition fact using tens to find each sum. The first three problems have been partially completed for you.

a. 4 + 8 = 10 + 2 = _____   f. 9 + 6 = ___ + ___ = _____
b. 5 + 9 = 10 + 4 = _____   g. 8 + 8 = ___ + ___ = _____
c. 7 + 4 = 10 + 1 = _____   h. 5 + 6 = ___ + ___ = _____
d. 8 + 5 = ___ + ___ = _____   i. 7 + 8 = ___ + ___ = _____
e. 6 + 4 = ___ + ___ = _____   j. 6 + 7 = ___ + ___ = _____

> You just reviewed information about finding sums.

24 Applying Addition

WHAT DO YOU KNOW ABOUT THE OPERATION OF ADDITION?

Think about the addition problem 4 + 12.

The problem has one addend that is a one-digit number.

The problem has a second addend that is a two-digit number.

The second addend is 12 and can be written as 10 + 2.

The sum of 4 + 12 can be solved as 4 + 10 + 2 or 10 + 6. The sum is 16.

▶ Using tens, find the sum of each addition problem. The first problem has been completed for you.

a. 6 + 13 can be solved as 6 + 10 + 3 or 10 + 9.
   The sum is 19.

b. 5 + 21 can be solved as 5 + ___ + ___ or 20 + 6.
   The sum is ___.

c. 7 + 32 can be solved as 7 + ___ + ___ or 30 + ___.
   The sum is ___.

d. 4 + 34 can be solved as ___ + ___ + ___ or 30 + ___.
   The sum is ___.

e. 45 + 2 can be solved as ___ + ___ + ___ or ___ + ___.
   The sum is ___.

f. 24 + 4 can be solved as ___ + ___ + ___ or ___ + ___.
   The sum is ___.

> You just reviewed how to use tens to find sums.

> Together, write four addition problems. Let one addend be a one-digit number and the second addend be a two-digit number. Use the method shown above to find each sum. Discuss each problem and its solution.

Applying Addition

PART TWO: Learn About Addition

Study the problem that Abby's teacher wrote on the chalkboard. As you study, think about the two ways that Abby solved the problem.

Problem: On Monday, students brought in 7 chocolate cakes for the school bake sale. On Tuesday, they brought in 6 more chocolate cakes. How many chocolate cakes did they bring in for the bake sale?

Two Ways That Abby Solved the Problem

1. I write the addends in any order. I use boxes to help me add. The order of the addends does not change the sum.

    $7 + 6 = 13$ ☐☐☐☐☐☐☐ + ☐☐☐☐☐☐

    $6 + 7 = 13$ ☐☐☐☐☐☐ + ☐☐☐☐☐☐☐

2. I use tens to find the sum.

    $7 + 6 = 10 + 3 = 13$ $6 + 7 = 10 + 3 = 13$

Answer: They have 13 chocolate cakes.

When you add, you put two or more numbers together to find the total, or sum. The numbers that you add together are called addends.

Three or more addends can be grouped in different ways.

$5 + 8 + 3 = 16$ $5 + 8 + 3 = 16$ $5 + 3 + 8 = 16$
$13 + 3 = 16$ $5 + 11 = 16$ $8 + 8 = 16$

You use **addition** to find the sum of two or more addends.

▶ Write addends in any order and get the same sum.

▶ Use tens to find the sum.

▶ When adding three or more numbers, group addends in different ways and get the same sum.

Applying Addition

Study another one of Abby's problems. Look at the two ways that Abby solved the problem. Then do Numbers 1 through 4.

Problem: Mr. Kelly sold 9 cans of yellow paint in the morning. He sold 8 more cans of yellow paint in the afternoon. How many cans of yellow paint did Mr. Kelly sell all together?

Use any order.	Use tens.
9 + 8 = 17 8 + 9 = 17	9 + 8 = 10 + 7 = 17

Answer: Mr. Kelly sold 17 cans of yellow paint.

1. Abby used any order and she used tens to find the sum of 7 and 8. Which of these does *not* show how Abby found the sum?
   - Ⓐ 7 + 8 = 15
   - Ⓑ 8 + 7 = 15
   - Ⓒ 9 + 6 = 14 + 1 = 15
   - Ⓓ 7 + 8 = 10 + 5 = 15

2. Abby's favorite CD has 12 songs on Disc 1 and 11 songs on Disc 2. What is the total number of songs on the two discs?
   - Ⓐ 16 songs
   - Ⓑ 14 songs
   - Ⓒ 26 songs
   - Ⓓ 23 songs

3. On her way to school, Abby counted 13 red traffic lights, 4 yellow lights, and 2 green lights. How many lights did she count in all?
   - Ⓐ 19 lights
   - Ⓑ 25 lights
   - Ⓒ 16 lights
   - Ⓓ 21 lights

4. Abby's dad ran 21 miles today and 15 miles yesterday. What is the total number of miles that he ran in two days?
   - Ⓐ 27 miles
   - Ⓑ 36 miles
   - Ⓒ 24 miles
   - Ⓓ 33 miles

Talk about your answers to questions 1–4. Tell why you chose the answers you did.

Applying Addition

PART THREE: Check Your Understanding

Remember: You use addition to find the sum of two or more addends.

▶ Write addends in any order, and get the same sum.

▶ Use tens to find the sum.

▶ When adding three or more numbers, group addends in different ways, and get the same sum.

Solve this problem. As you work, ask yourself, "How can I use the order of the addends? How can I use tens?"

5. Today, Abby mailed 12 invitations to her mother's surprise birthday party. Yesterday, she mailed 11 invitations. How many invitations did Abby mail in the two days?
   Ⓐ 23 invitations
   Ⓑ 19 invitations
   Ⓒ 26 invitations
   Ⓓ 17 invitations

Solve another problem. As you work, ask yourself, "How can I use grouping addends?"

6. Abby had 21 postcards. Recently she bought 13 postcards at a yard sale and 14 postcards at a gift shop. How many postcards does Abby now have?
   Ⓐ 35 postcards
   Ⓑ 48 postcards
   Ⓒ 39 postcards
   Ⓓ 40 postcards

Applying Addition

**Look at the answer choices for each question.
Read why each answer choice is correct or not correct.**

5. Today, Abby mailed 12 invitations to her mother's surprise birthday party. Yesterday, she mailed 11 invitations. How many invitations did Abby mail in the two days?

   ● 23 invitations

   This answer is correct because 12 + 11 = 23. You can also write 11 + 12 = 23. Changing the order of the addends does not change the sum.

   Ⓑ 19 invitations

   This answer is not correct because 12 + 11 does not equal 19. You can use tens to see that 12 + 11 = 10 + 10 + 3 = 23.

   Ⓒ 26 invitations

   This answer is not correct because 12 + 11 does not equal 26. You can use tens to see that 12 + 11 = 10 + 10 + 3 = 23.

   Ⓓ 17 invitations

   This answer is not correct because 12 + 11 does not equal 17. You can use tens to see that 12 + 11 = 10 + 10 + 3 = 23.

6. Abby had 21 postcards. Recently she bought 13 postcards at a yard sale and 14 postcards at a gift shop. How many postcards does Abby now have?

   Ⓐ 35 postcards

   This answer is not correct because 21 + 13 + 14 does not equal 35. Grouping addends can help get the correct answer: 21 + 13 = 34; and 34 + 14 = 48.

   ● 48 postcards

   This answer is correct because the sum of 21 + 13 + 14 is 48.

   Ⓒ 39 postcards

   This answer is not correct because 21 + 13 + 14 does not equal 39. Grouping addends can help get the correct answer: 21 + 13 = 34; and 34 + 14 = 48.

   Ⓓ 40 postcards

   This answer is not correct because 21 + 13 + 14 does not equal 40. Grouping addends can help get the correct answer: 21 + 13 = 34; and 34 + 14 = 48.

Applying Addition

PART FOUR: Learn More About Addition

You use addition to find the sum of addends. When an addend has three or more digits, think about the value of each digit.

▶ When you write the addends, line up the ones, tens, and hundreds in columns.

Add each column, from right to left, starting with the ones column.

If the sum of the ones is more than 10 ones, regroup 10 of the ones as 1 ten. Write 1 in the tens column to stand for the 10 ones you regrouped.

If the sum of the tens is more than 10 tens, regroup 10 of the tens as 1 hundred. Write 1 in the hundreds column.

hundreds	tens	ones
1	1	
1	3	5
+ 1	8	7
3	2	2

Abby helps out at Mr. Capra's grocery store. Do Numbers 7 through 10.

7. Mr. Capra sold 167 gallons of milk one day and 148 gallons the next day. How many gallons of milk did he sell in two days?
   - Ⓐ 215 gallons
   - Ⓑ 325 gallons
   - Ⓒ 315 gallons
   - Ⓓ 298 gallons

8. Abby and Mr. Capra counted 104 cans on one shelf and 97 cans on another shelf. How many cans did they count all together?
   - Ⓐ 201 cans
   - Ⓑ 191 cans
   - Ⓒ 218 cans
   - Ⓓ 174 cans

9. Abby bought a loaf of bread for $1.79 and a jar of peanut butter for $2.95. How much did Abby spend?
   - Ⓐ $4.64
   - Ⓑ $3.74
   - Ⓒ $3.75
   - Ⓓ $4.74

10. Mr. Capra paid Abby $5.75 to straighten the cans on the shelves and $3.75 to sweep the floor. How much did Mr. Capra pay Abby for her help?
    - Ⓐ $8.75
    - Ⓑ $9.50
    - Ⓒ $8.50
    - Ⓓ $9.00

Applying Addition

Read the journal entry that Abby wrote about her visit to her grandmother's house. Then do Numbers 11 through 14.

Saturday, May 13

Grandma and I got up early today and went on a nature hike in the woods behind her house. We followed deer tracks, but we didn't see any deer. We spent most of the morning watching for birds.

11. Abby's grandmother drove 125 miles to the city to pick up Abby. The drive back was 136 miles because she took a different route. How many miles did she drive to and from the city?

    Ⓐ 261 miles
    Ⓑ 157 miles
    Ⓒ 361 miles
    Ⓓ 251 miles

12. Abby and her grandmother walked into town to have lunch. Abby's lunch bill was $4.78. Grandma's lunch bill was $3.64. What was the sum of the two bills?

    Ⓐ $8.34
    Ⓑ $7.32
    Ⓒ $8.42
    Ⓓ $9.42

13. There are only 347 people in the town where Abby's grandmother lives. The neighboring town has 383 people. What is the total number of people in the two towns?

    Ⓐ 840 people
    Ⓑ 740 people
    Ⓒ 628 people
    Ⓓ 730 people

14. At a garage sale, Abby bought a pair of binoculars for $7.98 and a bird book for $2.65. How much did Abby spend at the garage sale?

    Ⓐ $10.63
    Ⓑ $9.76
    Ⓒ $11.73
    Ⓓ $10.73

Applying Addition

PART FIVE: Prepare for a Test

- A test question about addition may ask for the sum of two or more addends.
- A test question about addition may ask for the sum of numbers with two or three digits.

Read what Abby learned in a book about bats. Then do Numbers 15 and 16.

Bat News

There are over 900 different kinds of bats. Bats hear sounds that people can't hear. Bats use these sounds to find their way around in the dark and to hunt for insects at night.

Applying Addition

15. Abby read that there are 290 kinds of common bats and 154 kinds of flying fox bats and large fruit bats. What is the total number of the three kinds of bats?

    Ⓐ 544 kinds
    Ⓑ 394 kinds
    Ⓒ 345 kinds
    Ⓓ 444 kinds

Applying Addition

16. The list shows how many pictures of bats are in the book that Abby read.

    page 1   2 pictures
    page 3   8 pictures
    page 5   7 pictures

    How many pictures of bats are in the book?

    Ⓐ 17 pictures
    Ⓑ 25 pictures
    Ⓒ 13 pictures
    Ⓓ 27 pictures

Read about Abby's trip to a national park. Then do Numbers 17 and 18.

A Visit to Carlsbad Caverns

Last summer, Abby's grandmother took her to Carlsbad Caverns National Park, in New Mexico, to see Bat Cave. In summer, millions of bats come to live in the cool cave. Abby and her grandmother also took a tour of the park's underground caverns.

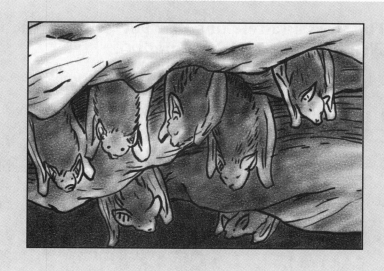

Applying Addition

17. Abby and her grandmother rode an elevator to the underground rooms beneath the park. They rode down 750 feet to the first level. They rode down 79 more feet to the second level. How many feet below the ground did the elevator take Abby and her grandmother?
    - Ⓐ 729 feet
    - Ⓑ 739 feet
    - Ⓒ 829 feet
    - Ⓓ 839 feet

Applying Addition

18. After their tour of Carlsbad Caverns, Abby and her grandmother visited the gift shop. Abby bought a book for $4.95, a bookmark for $2.00, and a postcard for 76¢. How much did Abby spend in the gift shop?
    - Ⓐ $6.93
    - Ⓑ $7.71
    - Ⓒ $8.61
    - Ⓓ $8.71

Strategies One–Three REVIEW

PART ONE: Read a Story

The story is a retelling of "The Three Little Pigs." Read the story. Then do Numbers 1 through 6.

Once, there were three pigs whose mud house was washed away in a rainstorm. "We should build our new house of bricks," the oldest and wisest pig said. "A brick house is sturdy. Best of all, there isn't a wolf anywhere who can blow down a brick house."

The lazy middle pig frowned. "Bricks are too heavy," he whined. "How about twigs? Twigs are nice and light," he said. "I vote for a twig house."

"The price of twigs has gone sky high!" the youngest pig squealed. "And bricks are WAY too expensive! How about straw?" he asked. "I vote for a straw house."

But, the pigs could not agree. So, the youngest pig built a straw house with 2 bundles of straw that cost a total of $2.31. He worked for 1 hour. It took 4 hours longer for the middle pig to build his house of 783 twigs that cost $7.12. The oldest pig worked for 9 hours more than the middle pig to build his house. He used about 3,000 bricks. He also emptied out his savings account of $27.86.

The next day, the tired old wolf saw the straw house and thought, "I can blow *that* down." So he did, with 9 huffs and 8 puffs.

The wolf chuckled when he saw the twig house. "I can blow *this* house

down, too," he said. So he did, with 17 huffs and 16 puffs.

The two pigs ran squealing to their older brother's brick house. "Let us in," they hollered. "The strongest, biggest wolf you've ever seen is after us!"

The wolf sighed when he saw the brick house. "I've never blown down a *brick* house," he said, "and I'm too old to try now." So the wolf just turned his tail and went home.

The three pigs lived happily ever after in the brick house.

Building Number Sense

1. The exact number of bricks that the oldest pig used to build his house is an odd number. Which of these could be the exact number of bricks?
   - Ⓐ 3,384 bricks
   - Ⓑ 3,101 bricks
   - Ⓒ 3,574 bricks
   - Ⓓ 3,028 bricks

Using Estimation

4. To the nearest 10¢, how much did the youngest pig and the middle pig spend to build both their houses?
   - Ⓐ $9.60
   - Ⓑ $9.10
   - Ⓒ $9.50
   - Ⓓ $9.40

Building Number Sense

2. If the middle pig had bought 1 more twig to build his house, how many twigs would he have had?
   - Ⓐ 790 twigs
   - Ⓑ 780 twigs
   - Ⓒ 782 twigs
   - Ⓓ 784 twigs

Applying Addition

5. How long did it take the oldest pig to build his brick house?
   - Ⓐ 11 hours
   - Ⓑ 12 hours
   - Ⓒ 13 hours
   - Ⓓ 14 hours

Using Estimation

3. To the nearest 10¢, how much money did the oldest pig spend to build his house?
   - Ⓐ $27.90
   - Ⓑ $27.60
   - Ⓒ $30.00
   - Ⓓ $27.80

Applying Addition

6. What was the total number of huffs and puffs that the wolf took to blow down the straw house and the twig house?
   - Ⓐ 50 huffs and puffs
   - Ⓑ 48 huffs and puffs
   - Ⓒ 45 huffs and puffs
   - Ⓓ 52 huffs and puffs

PART TWO: Read an Article

Latoya read an article about desert plants. Read the article. Then do Numbers 7 through 12.

Desert plants have special ways to stay alive in the hot, dry climate of the desert. One way is by dropping their seeds on the desert floor so that new plants will grow. Another way that desert plants survive is by saving water.

The cactus is a desert plant that has a thick, fleshy trunk. When rain falls in the desert, the roots of the cactus drink in water and store it in the cactus's trunk.

One of the largest cacti in the world is the saguaro (pronounced *sah-wah´-ro*). The saguaro grows only in the Sonoran Desert, in the southwestern part of the United States. The Sonoran Desert gets less than 12 inches of rain a year, but the roots of the saguaro are not deep. So, even when it rains for only a few minutes, the saguaro can get water.

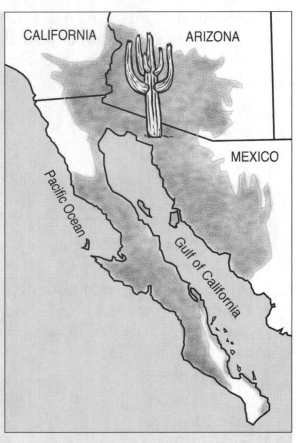

As it drinks in water, the pleated trunk of the saguaro puffs out. In a good rain, an adult saguaro can drink in 200 gallons of water. This amount of water can last a whole year.

The saguaro grows to be a giant, but it doesn't grow very quickly. After a year, a saguaro seedling is less than 1 inch tall. It may grow only 11 more inches in the next 15 years.

When a saguaro is about 30 years old, it gets flowers. These creamy-white flowers bloom at night and live for only 24 hours. After the flowers blossom, a small fruit appears. A single saguaro fruit can contain 2,000 seeds. The pulp and seeds of the ripened fruit provide food for many desert creatures.

At about 75 years of age, a saguaro is around 15 feet tall. That's when it begins to get branches that look like bent arms. A saguaro that is 150 years old can be 50 feet tall and have dozens of arms.

Building Number Sense

7. Latoya thinks that the shaded cactus in the row is the prettiest. In what place is the shaded cactus?

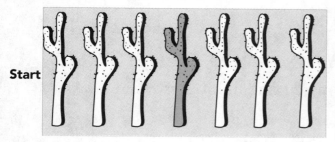

Ⓐ third Ⓒ sixth
Ⓑ seventh Ⓓ fourth

Building Number Sense

8. Latoya read that the exact number of seeds in one saguaro fruit was 1,974. What is the value of the 9 in 1,974?

Ⓐ 9
Ⓑ 90
Ⓒ 900
Ⓓ 9,000

Using Estimation

9. Latoya learned that Saguaro National Park, a part of the Sonoran Desert, covers 87,114 acres. Which of these is 87,114 rounded to the nearest thousand?

Ⓐ 87,000
Ⓑ 87,100
Ⓒ 87,150
Ⓓ 7,000

Using Estimation

10. Latoya estimated, in inches, the sum of the heights of three saguaros. To the nearest hundred inches, what would the sum be?

Cactus 1   528 inches
Cactus 2   600 inches
Cactus 3   516 inches

Ⓐ 1,650 inches
Ⓑ 1,700 inches
Ⓒ 1,600 inches
Ⓓ 1,500 inches

Applying Addition

11. Latoya added the ages of two saguaros. One was 78 years old, and the other was 161 years old. What is the sum of their ages?

Ⓐ 841 years
Ⓑ 239 years
Ⓒ 139 years
Ⓓ 247 years

Applying Addition

12. During one rainy summer, a saguaro soaked up 145 gallons of water one day and 62 gallons the next day. How many gallons of water did the saguaro soak up in the two days?

Ⓐ 107 gallons
Ⓑ 765 gallons
Ⓒ 217 gallons
Ⓓ 207 gallons

Strategies 1–3 Review

Strategy Four: APPLYING SUBTRACTION

PART ONE: Think About Subtraction

WHAT DO YOU KNOW ABOUT SUBTRACTION?

Subtraction is one of the four basic operations.

Subtraction is used to find how much is left when one number is taken away from another number.

Subtraction is performed on two numbers. In the problem $12 - 5 = 7$, 12 is the *minuend*, 5 is the *subtrahend*, and 7 is the answer, or *difference*.

To do subtraction correctly, you must know the basic subtraction facts.

Subtraction facts are related to addition facts in fact families. For example, $7 + 9 = 16$, $9 + 7 = 16$, $16 - 9 = 7$, and $16 - 7 = 9$ are a fact family.

▶ What two subtraction facts are related to each addition fact? The first and second set of facts have been completed for you.

a. $8 + 4 = 12$ d. $8 + 6 = 14$ g. $9 + 6 = 15$ j. $7 + 3 = 10$
   $12 - 8 = \underline{\ 4\ }$ _____ _____ _____
   $12 - 4 = \underline{\ 8\ }$ _____ _____ _____

b. $7 + 6 = 13$ e. $5 + 2 = 7$ h. $6 + 4 = 10$ k. $9 + 8 = 17$
   $13 - 7 = \underline{\ 6\ }$ _____ _____ _____
   $13 - 6 = \underline{\ 7\ }$ _____ _____ _____

c. $8 + 7 = 15$ f. $9 + 2 = 11$ i. $4 + 3 = 7$ l. $9 + 5 = 14$
   _____ _____ _____ _____
   _____ _____ _____ _____

▶ How many related subtraction facts are there when a number is added to itself? Write two examples to show your answer.

_____

_____

> You just reviewed information about how subtraction is related to addition.

Applying Subtraction

What Do You Know About Regrouping in Subtraction?

Look at the subtraction problem 22 − 7.

The minuend is 22 or 2 tens + 2 ones.

The subtrahend is 7 or 7 ones.

You cannot subtract 7 ones from 2 ones.

You must regroup one of the tens, and change it to ones. One ten becomes 10 ones.

Add 10 ones to 2 ones to get 12 ones.

Subtract 7 ones from 12 ones to get 5 ones.

Subtract 0 tens from 1 ten to get 1 ten.

The difference in the problem 22 − 7 is 15, or 1 ten and 5 ones.

▶ Study the same subtraction problem, 22 − 7 = 15, as it is regrouped below. Then solve the subtraction problems that follow.

tens	ones
$\overset{1}{\cancel{2}}$	$\overset{12}{\cancel{2}}$
−	7
1	5

a. 26
  − 7

b. 28
  −19

c. 46
  −44

d. 34
  − 8

e. 37
  −28

f. 33
  −24

g. 28
  − 7

h. 24
  −14

i. 21
  −16

j. 42
  − 6

k. 45
  −23

l. 37
  −23

You just reviewed information about regrouping in subtraction.

Together, write four subtraction problems. Let each minuend be a two-digit number and each subtrahend be a one-digit or two-digit number. Then solve the problems. Regroup as needed. Discuss each problem and its solution.

Applying Subtraction

PART TWO: Learn About Subtraction

Study the problem that Pablo's teacher wrote on the chalkboard.
As you study, think about the two ways that Pablo solved the problem.

Problem: A box has 14 pencils.
Each of the 8 children in a group takes one pencil.
How many pencils are left in the box?

Two Ways That Pablo Solved the Problem

1. I write the numbers across. I write the larger number first. I subtract the smaller number from the larger number. I use boxes to help me subtract.

   $14 - 8 = 6$
   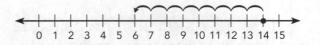

2. I use a number line to count 8 back from 14 to find the answer.

Answer: There are 6 pencils left in the box.

When you subtract, you find the difference between two numbers. The larger number is the minuend. The smaller number is the subtrahend. The difference, or answer, is always smaller than the minuend, unless you subtract 0.

You use subtraction to find out how many are left or how many more are needed.

You can write a subtraction problem in a row. Put the larger number or minuend first. You can write a subtraction problem in a column. Put the larger number or minuend on top.

$23 - 11 = 12$ 23 minuend
                 -11 subtrahend
                  12 difference

You use **subtraction** to find the difference between two numbers.

▶ Write numbers in a subtraction problem in a row or in a column.

▶ Count back from the larger number, or minuend, to find the difference.

Applying Subtraction

Study another one of Pablo's problems. Look at the two ways that Pablo solved the problem. Then do Numbers 1 through 4.

Problem: Last year, 15 new houses were built in Oakville.
This year, 7 new houses were built in Oakville.
How many more new houses were built last year?

Use boxes to count. Take away or cross out.	Count back. Use a number line.
▫▫▫▫▫▫▫ ▫▫▫▫▫▫▫▫ 15 − 7 = 8	0 1 2 3 4 5 6 7 8 9 10 11 12 13 14 15 16 15 − 7 = 8

Answer: Last year, 8 more new houses were built.

1. There were 18 birds in the large oak tree outside Pablo's house. Suddenly, 6 birds flew away. How many birds stayed in the tree?
   - Ⓐ 12 birds
   - Ⓑ 16 birds
   - Ⓒ 5 birds
   - Ⓓ 8 birds

2. Pablo saw 11 ducks by the edge of a pond. Soon 7 of the ducks swam away. How many ducks remained by the edge of the pond?
   - Ⓐ 8 ducks
   - Ⓑ 9 ducks
   - Ⓒ 2 ducks
   - Ⓓ 4 ducks

3. On a visit to his uncle's farm, Pablo counted 12 pigs and 10 cows. How many more pigs than cows were on the farm?
   - Ⓐ 12 pigs
   - Ⓑ 8 pigs
   - Ⓒ 4 pigs
   - Ⓓ 2 pigs

4. Pablo counted 17 ladybugs and 7 grasshoppers in his family's flower garden. How many more ladybugs than grasshoppers did he count?
   - Ⓐ 9 ladybugs
   - Ⓑ 15 ladybugs
   - Ⓒ 1 ladybug
   - Ⓓ 10 ladybugs

Talk about your answers to questions 1–4. Tell why you chose the answers you did.

Applying Subtraction

PART THREE: Check Your Understanding

Remember: You use subtraction to find the difference between two numbers.

▶ Write numbers in a subtraction problem in a row or in a column.

▶ Count back from the larger number, or minuend, to find the difference.

Solve this problem. As you work, ask yourself, "What number is the larger number? What number do I take away?"

5. Pablo and his sister Maria were measured by their doctor. Pablo is 56 inches tall. Maria is 51 inches tall. How much taller is Pablo than Maria?
   - Ⓐ 15 inches
   - Ⓑ 4 inches
   - Ⓒ 17 inches
   - Ⓓ 5 inches

Solve another problem. As you work, ask yourself, "How can I use counting back to help me find the difference?"

6. On Tuesday, Pablo's doctor saw a total of 19 children in her office. In the morning, the doctor saw 8 children. How many children did she see in the afternoon?
   - Ⓐ 11 children
   - Ⓑ 10 children
   - Ⓒ 9 children
   - Ⓓ 1 child

Applying Subtraction

Look at the answer choices for each question.
Read why each answer choice is correct or not correct.

5. Pablo and his sister Maria were measured by their doctor. Pablo is 56 inches tall. Maria is 51 inches tall. How much taller is Pablo than Maria?

   Ⓐ 15 inches

   This answer is not correct because if you take 51 away from 56, you have 5 left, not 15.

   Ⓑ 4 inches

   This answer is not correct because if you take 51 away from 56, you have 5 left, not 4.

   Ⓒ 17 inches

   This answer is not correct because if you take 51 away from 56, you have 5 left, not 17.

   ● 5 inches

   This answer is correct because if you take 51 away from 56, you have 5 left. So, 56 − 51 = 5.

6. On Tuesday, Pablo's doctor saw a total of 19 children in her office. In the morning, the doctor saw 8 children. How many children did she see in the afternoon?

   ● 11 children

   This answer is correct because if you count back 8 from 19, you get 11.

   Ⓑ 10 children

   This answer is not correct because if you count back 8 from 19, you get 11, not 10.

   Ⓒ 9 children

   This answer is not correct because if you count back 8 from 19, you get 11, not 9.

   Ⓓ 1 child

   This answer is not correct because if you count back 8 from 19, you get 11, not 1.

Applying Subtraction

PART FOUR: Learn More About Subtraction

You use subtraction to find the difference between two numbers. When one or both of the numbers have two or more digits, think about the value of each digit.

▶ When you subtract, line up the ones, tens, and hundreds columns. When you subtract money, line up the decimal points. Subtract each column from right to left.

Look at the ones column. If the number you are taking away is larger than the number you are taking away from, regroup 1 ten to get 10 ones.

Look at the tens column. If necessary, regroup 1 hundred to get 10 tens.

Find the difference. Subtract the ones column. Then subtract the tens column. Finally, subtract the hundreds column.

hundreds	tens	ones
4	11	10
5̸	2̸	0̸
− 3	5	4
1	6	6

Pablo entered a neighborhood race. Do Numbers 7 through 10.

7. A total of 247 people signed up for the Oakville Fun Run. On race day, 169 people signed up to run. How many people signed up before race day?
   - Ⓐ 122 people
   - Ⓑ 82 people
   - Ⓒ 88 people
   - Ⓓ 78 people

8. Each child paid $5.75 to enter the race. Each adult paid $15.50. How much less did a child pay to enter the race than an adult?
   - Ⓐ $9.75
   - Ⓑ $10.25
   - Ⓒ $9.85
   - Ⓓ $10.75

9. Pablo completed the children's race in 65 minutes. His dad ran farther and longer. It took him 131 minutes. How many minutes more than Pablo did his dad run?
   - Ⓐ 74 minutes
   - Ⓑ 66 minutes
   - Ⓒ 76 minutes
   - Ⓓ 61 minutes

10. Members of the Fun Run Committee ordered 864 Fun Run T-shirts. They sold 578 of them. How many T-shirts were *not* sold?
    - Ⓐ 314 T-shirts
    - Ⓑ 376 T-shirts
    - Ⓒ 286 T-shirts
    - Ⓓ 216 T-shirts

Read what Pablo wrote about ant farms. Then do Numbers 11 through 14.

I have two ant farms. The first one I built with my mom's help. We used ants from my backyard. The second ant farm was a gift from my aunt on my birthday. I think ants make great pets. They are easy to care for and fun to watch. The most important ant in an ant farm is the queen ant. She lays the eggs. The "soldier" ants are the workers. They collect and gather food.

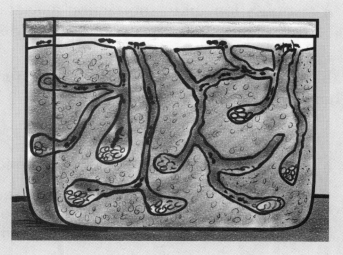

11. Pablo's mom took him shopping for materials to build his first ant farm. The bill at the hardware store was $7.54. Pablo gave $10.00 to the cashier. How much change did he get back?
Ⓐ $3.54
Ⓑ $2.44
Ⓒ $3.56
Ⓓ $2.46

12. Pablo spent a total of $9.81 for materials to build his own ant farm. Pablo's aunt paid $22.35 for the ant farm at Natureland. How much more did his aunt spend for the store-bought ant farm?
Ⓐ $13.56
Ⓑ $13.44
Ⓒ $12.54
Ⓓ $12.16

13. Pablo gathered 123 ants from his backyard. He put 36 of them into the ant farm he built. He gave the rest to friends who were starting their own ant farms. How many ants did he give to his friends?
Ⓐ 87 ants
Ⓑ 143 ants
Ⓒ 109 ants
Ⓓ 47 ants

14. The ants in Pablo's first ant farm dug tunnels that were a total of 282 centimeters long. The ants in Pablo's second ant farm dug tunnels that were a total of 401 centimeters long. How much longer were the tunnels in the second ant farm?
Ⓐ 211 cm
Ⓑ 119 cm
Ⓒ 129 cm
Ⓓ 181 cm

Applying Subtraction

PART FIVE: Prepare for a Test

▶ A test question about subtraction may ask for the difference between two numbers.

▶ A test question about subtraction may ask for the difference between numbers with two or more digits.

Pablo read an article about a home-run race. Read the article. Then do Numbers 15 and 16.

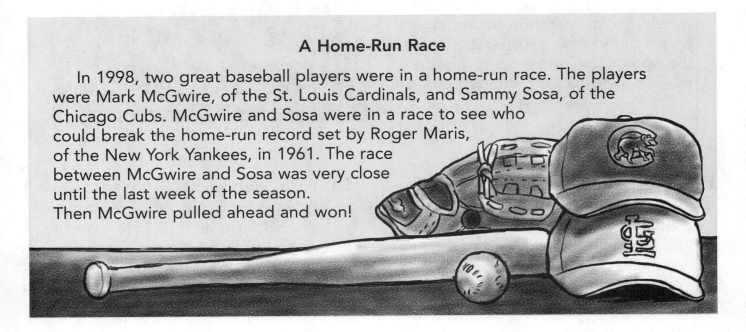

A Home-Run Race

In 1998, two great baseball players were in a home-run race. The players were Mark McGwire, of the St. Louis Cardinals, and Sammy Sosa, of the Chicago Cubs. McGwire and Sosa were in a race to see who could break the home-run record set by Roger Maris, of the New York Yankees, in 1961. The race between McGwire and Sosa was very close until the last week of the season. Then McGwire pulled ahead and won!

Applying Subtraction

15. Mark McGwire and Sammy Sosa hit a total of 136 home runs in the 1998 season. Mark McGwire hit 70 home runs. How many home runs did Sammy Sosa hit?

    Ⓐ 76 home runs
    Ⓑ 64 home runs
    Ⓒ 66 home runs
    Ⓓ 72 home runs

Applying Subtraction

16. Mark McGwire hit 70 home runs. Roger Maris hit 61 home runs in 1961. How many more home runs did Mark McGwire hit than Roger Maris to set a new record?

    Ⓐ 9 home runs
    Ⓑ 11 home runs
    Ⓒ 1 home run
    Ⓓ 10 home runs

Read what Pablo learned about Chicago's famous ballpark.
Then do Numbers 17 and 18.

Pablo's First Pro Game

Pablo's parents took him to a baseball game at Wrigley Field, in Chicago. Before the game, Pablo and his dad read about Wrigley Field. They learned that the park was built in 1914 for a team called the Chicago Federals. Later, it became the home field of the Chicago Cubs and was called Cubs Park. In 1926, the park was renamed Wrigley Field, for William Wrigley, Jr., the owner of the Cubs.

Pablo also learned two baseball customs that started at Wrigley Field: letting fans keep foul balls hit into the stands, and the singing of "The Star-Spangled Banner" before a game.

Applying Subtraction

17. Pablo learned that no baseball player has ever hit a ball to the scoreboard in center field. The distance from home plate to center field is 400 feet. The distance to right field is 353 feet. How much farther is it to center field than to right field?
    - Ⓐ 49 feet
    - Ⓑ 53 feet
    - Ⓒ 57 feet
    - Ⓓ 47 feet

Applying Subtraction

18. Pablo's parents paid $11.00 for each of their tickets in the family section of the stadium. Tickets for the bleachers section are $6.00 each. How much more does a ticket in the family section cost?
    - Ⓐ $10.40
    - Ⓑ $5.00
    - Ⓒ $4.00
    - Ⓓ $7.50

Applying Subtraction

Strategy Five: APPLYING MULTIPLICATION

PART ONE: Think About Multiplication

WHAT DO YOU KNOW ABOUT MULTIPLICATION?

Multiplication is one of the four basic operations.

Multiplication is repeated addition: 4×2 can mean four 2's, or $2 + 2 + 2 + 2$.

The numbers in a multiplication problem are called *factors*. The factors are 4 and 2.

The answer in multiplication is called the *product*. The product of 4×2 is 8.

To do multiplication correctly, you must know the basic multiplication facts.

▶ Write each repeated addition problem as a multiplication problem and solve it.

a. $3 + 3 + 3 + 3 =$ _____
b. $6 + 6 + 6 + 6 + 6 =$ _____
c. $7 + 7 + 7 + 7 + 7 + 7 =$ _____
d. $5 + 5 + 5 =$ _____

e. $2 + 2 + 2 + 2 + 2 + 2 =$ _____
f. $4 + 4 + 4 + 4 + 4 + 4 + 4 =$ _____
g. $9 + 9 + 9 + 9 + 9 + 9 =$ _____
h. $8 + 8 + 8 + 8 + 8 =$ _____

> You just reviewed how multiplication is related to addition.

▶ Write the product for each basic multiplication fact.

a. $6 \times 7 =$ ___
$7 \times 6 =$ ___

b. $8 \times 9 =$ ___
$9 \times 8 =$ ___

c. $7 \times 9 =$ ___
$9 \times 7 =$ ___

d. $3 \times 9 =$ ___
$9 \times 3 =$ ___

e. $5 \times 6 =$ ___
$6 \times 5 =$ ___

f. $9 \times 4 =$ ___
$4 \times 9 =$ ___

g. $5 \times 7 =$ ___
$7 \times 5 =$ ___

h. $3 \times 8 =$ ___
$8 \times 3 =$ ___

▶ What do you notice about each pair of problems above?

Changing the order of the _____ does not change the _____.

> You just reviewed information about basic facts and the order of factors in multiplication.

WHAT DO YOU KNOW ABOUT THE OPERATION OF MULTIPLICATION?

Look at the multiplication problem 3 × 14.

The problem has one factor that is a one-digit number.

The problem has a second factor that is a two-digit number.

The second factor is 14 and can be written as 10 + 4.

So, 3 × 14 can be solved as (3 × 10) + (3 × 4) or 30 + 12.
The product of 3 × 14 is 42.

▶ Using tens, what is the product of each multiplication problem? The first problem has been completed for you.

  a. 6 × 12 can be solved as (6 × 10) + (6 × 2) or 60 + 12.
  The product of 6 × 12 is 72.

  b. 5 × 21 can be solved as (5 × __) + (5 × __) or ____ + ____.
  The product of 5 × 21 is _____.

  c. 7 × 34 can be solved as (__ × __) + (__ × __) or ____ + ____.
  The product of 7 × 34 is _____.

  d. 4 × 57 = (__ × __) + (__ × __) or ____ + ____.
  The product of 4 × 57 is _____.

  e. 2 × 23 = (__ × __) + (__ × __) or ____ + ____.
  The product of 2 × 23 is _____.

  f. 3 × 56 = (__ × __) + (__ × __) or ____ + ____.
  The product of 3 × 56 is _____.

> You just reviewed how to use tens to find products.

> Together, write four multiplication problems. Each problem should have one factor that is a one-digit number and a second factor that is a two-digit number. Use the steps shown above to find each product.

Applying Multiplication

PART TWO: Learn About Multiplication

Study the problem that Lou's teacher wrote on the chalkboard.
As you study, think about the five ways that Lou solved the problem.

Problem: The students lined up for a class photo.
They stood in 4 rows.
There were 6 students in each row.
How many students lined up for the picture?

Five Ways That Lou Solved the Problem	
1. I draw a picture.	☐☐☐☐☐☐ ☐☐☐☐☐☐ ☐☐☐☐☐☐ ☐☐☐☐☐☐
2. I write a repeated addition sentence.	$6 + 6 + 6 + 6 = 24$
3. I write a different repeated addition sentence.	$4 + 4 + 4 + 4 + 4 + 4 = 24$
4. I write a multiplication sentence.	$4 \times 6 = 24$
5. I change the order to write a different multiplication sentence.	$6 \times 4 = 24$

Answer: There were 24 students lined up for the picture.

When you multiply, you use repeated addition to find a total.
The numbers that you multiply are called factors. The total is called a product.
You may change the order of the factors. The product does not change.

$$3 \times 6 = 18 \qquad 6 \times 3 = 18$$

You use **multiplication** to find the product of two factors.

▶ To help find a product, write a repeated addition sentence for the problem. Then write a multiplication sentence for the same problem.

▶ Changing the order of the factors does not change the product.

Applying Multiplication

Study another one of Lou's problems. Think about the two ways that Lou solved the problem. Then do Numbers 1 through 4.

Problem: In the garden, there are 5 rows of tomatoes.
There are 7 tomato plants in each row.
How many tomato plants are in the garden?

Write a repeated addition sentence.	$7 + 7 + 7 + 7 + 7 = 35$ OR $5 + 5 + 5 + 5 + 5 + 5 + 5 = 35$
Write a multiplication sentence.	$7 \times 5 = 35$ OR $5 \times 7 = 35$

Answer: There are 35 tomato plants in the garden.

1. Lou looked at his own garden. He saw 3 rows of carrots. There were 9 carrots in each row. How many carrots were there in all?
   Ⓐ 30 carrots
   Ⓑ 27 carrots
   Ⓒ 25 carrots
   Ⓓ 24 carrots

2. Lou and his 3 sisters dug potatoes from the garden. They each dug 6 potatoes. Together, how many potatoes did they dig?
   Ⓐ 24 potatoes
   Ⓑ 18 potatoes
   Ⓒ 28 potatoes
   Ⓓ 22 potatoes

3. Lou picked fresh peppers for dinner. He picked 4 peppers from each plant. There were 5 pepper plants. How many peppers did Lou pick?
   Ⓐ 10 peppers
   Ⓑ 25 peppers
   Ⓒ 15 peppers
   Ⓓ 20 peppers

4. Lou looked for ripe tomatoes. There were 4 plants with ripe tomatoes. Each plant had 3 ripe tomatoes. How many ripe tomatoes did Lou see?
   Ⓐ 7 ripe tomatoes
   Ⓑ 11 ripe tomatoes
   Ⓒ 12 ripe tomatoes
   Ⓓ 16 ripe tomatoes

Talk about your answers to questions 1–4. Tell why you chose the answers you did.

Applying Multiplication

PART THREE: Check Your Understanding

Remember: You use multiplication to find the product of two factors.

▶ To help find a product, write a repeated addition sentence for the problem. Then write a multiplication sentence for the same problem.

▶ Changing the order of the factors does not change the product.

Solve this problem. As you work, ask yourself, "How can I use repeated addition to help me find the product?"

5. Lou's soccer team won its first game. After the game, the 3 coaches took the entire team out for ice cream. Each coach drove 5 players. How many players are on Lou's soccer team?
   - Ⓐ 15 players
   - Ⓑ 12 players
   - Ⓒ 18 players
   - Ⓓ 14 players

Solve another problem. As you work, ask yourself, "What multiplication sentence do I write for this problem?"

6. At the end of the season, Lou's team wanted to buy gifts for the coaches. Each player gave 4 dollars to buy the gifts. Lou's mother collected the money. One week, she collected from 8 of the players. How much money did Lou's mom collect that week?
   - Ⓐ 12 dollars
   - Ⓑ 24 dollars
   - Ⓒ 32 dollars
   - Ⓓ 22 dollars

Applying Multiplication

Look at the answer choices for each question.
Read why each answer choice is correct or not correct.

5. Lou's soccer team won its first game. After the game, the 3 coaches took the entire team out for ice cream. Each coach drove 5 players. How many players are on Lou's soccer team?

   ● **15 players**

   This answer is correct because $5 + 5 + 5 = 15$.

   Ⓑ 12 players

   This answer is not correct because $5 + 5 + 5 = 15$, not 12. You may not have added correctly.

   Ⓒ 18 players

   This answer is not correct because $5 + 5 + 5 = 15$, not 18. You may not have added correctly.

   Ⓓ 14 players

   This answer is not correct because $5 + 5 + 5 = 15$, not 14. You may not have added correctly.

6. At the end of the season, Lou's team wanted to buy gifts for the coaches. Each player gave 4 dollars to buy the gifts. Lou's mother collected the money. One week, she collected from 8 of the players. How much money did Lou's mom collect that week?

   Ⓐ 12 dollars

   This answer is not correct because $8 \times 4 = 32$ is the correct multiplication sentence for $4 + 4 + 4 + 4 + 4 + 4 + 4 + 4 = 32$. You may have written $4 + 4 + 4 = 12$.

   Ⓑ 24 dollars

   This answer is not correct because $8 \times 4 = 32$ is the correct multiplication sentence for $4 + 4 + 4 + 4 + 4 + 4 + 4 + 4 = 32$. You may have written $4 + 4 + 4 + 4 + 4 + 4 = 24$.

   ● **32 dollars**

   This answer is correct because $8 \times 4 = 32$ is the correct multiplication sentence for $4 + 4 + 4 + 4 + 4 + 4 + 4 + 4 = 32$.

   Ⓓ 22 dollars

   This answer is not correct because $8 \times 4 = 32$ is the correct multiplication sentence for $4 + 4 + 4 + 4 + 4 + 4 + 4 + 4 = 32$. You may not have multiplied correctly.

Applying Multiplication

PART FOUR: Learn More About Multiplication

In some multiplication problems, one factor is a two-digit number. For these problems, write the factors in columns. Place the two-digit number at the top.

▶ First, multiply the ones. Write the product in the ones column. Next, multiply the tens. Write the product in the tens column.

tens	ones
1	2
×	4
	8

$4 \times 2 = 8$

tens	ones
1	2
×	4
4	8

$4 \times 10 = 40$
$12 \times 4 = 48$

▶ If the product of the ones is more than 10, regroup each 10 ones to get 1 ten. Above the tens column, write the number of 10 ones that were regrouped. Then multiply. Add the extra tens before writing the total tens.

Multiply the ones. Regroup into 8 ones and 2 tens. Write 8 in the ones column. Write 2 above the tens column.

tens	ones
2	
1	7
×	4
	8

$4 \times 7 = 28$

Multiply the tens. Add 2 tens. Write 6 in the tens column.

tens	ones
2	
1	7
×	4
6	8

$4 \times 10 = 40$
$40 + 20 = 60$
$17 \times 4 = 68$

Lou collects stamps. His friends wrote some math problems about the stamps. Do Numbers 7 through 10.

7. Lou's favorite stamps are each worth 13 cents. He has 6 of them. What is the total value of the stamps?
   - Ⓐ 72 cents
   - Ⓑ 78 cents
   - Ⓒ 88 cents
   - Ⓓ 68 cents

8. Lou has 15 stamps from 1917 and 1932. He will sell each stamp for 3 cents. What will be the selling price of all the stamps?
   - Ⓐ 45 cents
   - Ⓑ 50 cents
   - Ⓒ 40 cents
   - Ⓓ 35 cents

9. Lou bought 4 stamps. He paid 21 cents for each stamp. How much did Lou spend for the 4 stamps?
   - Ⓐ 94 cents
   - Ⓑ 90 cents
   - Ⓒ 84 cents
   - Ⓓ 88 cents

10. Lou ordered 3 stamps from a Web site. Each stamp is worth 29 cents. What is the value of the stamps that Lou ordered?
    - Ⓐ 92 cents
    - Ⓑ 77 cents
    - Ⓒ 84 cents
    - Ⓓ 87 cents

Applying Multiplication

Lou read a book about time. Read what Lou learned.
Then do Numbers 11 through 14.

There are many ways to measure time. You can measure it by seconds, minutes, or hours. For longer amounts of time, you can measure by days, weeks, months, or years. For very long amounts of time, you can use decades, centuries, or millenniums. For very small amounts of time, you can use nanoseconds. There are 1 billion nanoseconds in 1 second. Usually, only scientists measure with nanoseconds.

11. Lou knows that there are 12 months in a year. Lou correctly determined the number of months in 7 years. What was Lou's answer?
    Ⓐ 74 months
    Ⓑ 80 months
    Ⓒ 84 months
    Ⓓ 88 months

12. Lou reads for 19 hours every month. He correctly figured the number of hours he reads in 5 months. What was Lou's answer?
    Ⓐ 94 hours
    Ⓑ 85 hours
    Ⓒ 95 hours
    Ⓓ 98 hours

13. Usually, 4 days of each week in Lou's town are sunny. At that rate of sunny days per week, about how many sunny days would there be in 16 weeks?
    Ⓐ 64 sunny days
    Ⓑ 60 sunny days
    Ⓒ 54 sunny days
    Ⓓ 40 sunny days

14. Lou learned that there are 10 years in a decade. He correctly determined the number of years in 5 decades. What was Lou's answer?
    Ⓐ 45 years
    Ⓑ 60 years
    Ⓒ 55 years
    Ⓓ 50 years

PART FIVE: Prepare for a Test

▶ A test question about multiplication may ask for the product of two single-digit factors.

▶ A test question about multiplication may ask for the product of a single-digit factor and a two-digit factor.

Lou learned some facts about bones of the body. Read what Lou learned. Then do Numbers 15 and 16.

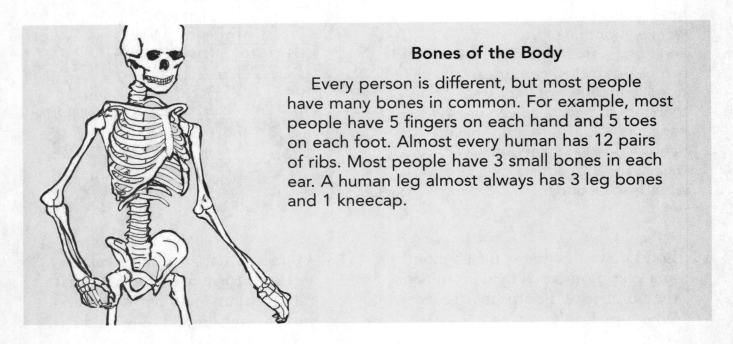

Bones of the Body

Every person is different, but most people have many bones in common. For example, most people have 5 fingers on each hand and 5 toes on each foot. Almost every human has 12 pairs of ribs. Most people have 3 small bones in each ear. A human leg almost always has 3 leg bones and 1 kneecap.

Applying Multiplication

15. Lou read that there are 4 bones in each human leg. Lou correctly figured how many bones there are in the legs of 4 people. What was Lou's answer?
    Ⓐ 28 bones
    Ⓑ 32 bones
    Ⓒ 16 bones
    Ⓓ 34 bones

Applying Multiplication

16. Lou learned that a person has 24 ribs. He correctly figured the number of ribs 2 people have. What was Lou's answer?
    Ⓐ 44 ribs
    Ⓑ 52 ribs
    Ⓒ 36 ribs
    Ⓓ 48 ribs

Applying Multiplication

Read this part of a report that Lou wrote about volcanoes. Then do Numbers 17 and 18.

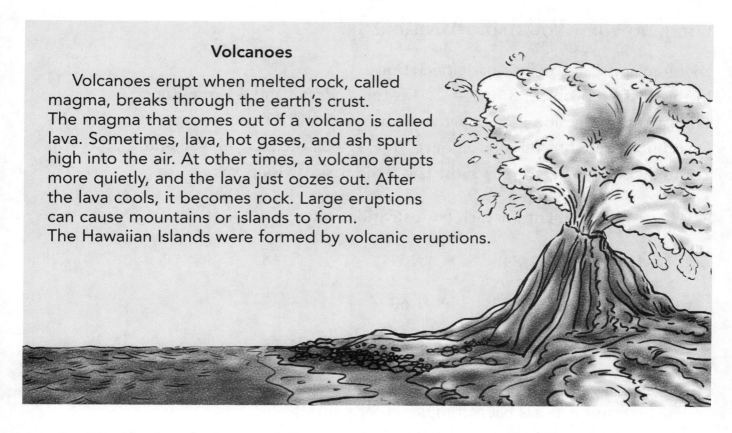

Volcanoes

Volcanoes erupt when melted rock, called magma, breaks through the earth's crust. The magma that comes out of a volcano is called lava. Sometimes, lava, hot gases, and ash spurt high into the air. At other times, a volcano erupts more quietly, and the lava just oozes out. After the lava cools, it becomes rock. Large eruptions can cause mountains or islands to form. The Hawaiian Islands were formed by volcanic eruptions.

Applying Multiplication

17. Lou read that three different volcanoes erupted a total of 14 times in 2004. If this number of eruptions occurs every year, how many eruptions of the three volcanoes would there be in 6 years?

   Ⓐ 84 eruptions
   Ⓑ 80 eruptions
   Ⓒ 89 eruptions
   Ⓓ 74 eruptions

Applying Multiplication

18. Lou also read that Stromboli, a volcano near Sicily, erupts 3 times every hour. Lou correctly figured how many times Stromboli erupts each day. What was Lou's answer?

   Ⓐ 64 times
   Ⓑ 72 times
   Ⓒ 75 times
   Ⓓ 67 times

Strategy Six APPLYING DIVISION

PART ONE: Think About Division

WHAT DO YOU KNOW ABOUT DIVISION?

Division is one of the four basic operations.
In the division problem $8 \div 4 = 2$, 8 is the *dividend* and 4 is the *divisor*.
The answer in division is the *quotient*. The quotient is 2.
Division is related to multiplication: if $8 \div 4 = 2$, then $2 \times 4 = 8$ and $4 \times 2 = 8$.
To do division correctly, you must know the basic division facts.

▶ What is the quotient for each basic division fact?

a. $6\overline{)42}$ d. $4\overline{)24}$ g. $8\overline{)64}$ j. $8\overline{)48}$

b. $9\overline{)27}$ e. $5\overline{)30}$ h. $9\overline{)63}$ k. $5\overline{)35}$

c. $8\overline{)72}$ f. $7\overline{)56}$ i. $6\overline{)54}$ l. $4\overline{)36}$

▶ Write a pair of division facts that is related to each pair of multiplication facts. The first problem has been completed for you.

a. $4 \times 8 = 32$ c. $9 \times 5 =$ ___   e. $7 \times 9 =$ ___   g. $6 \times 3 =$ ___
   $8 \times 4 = 32$ $5 \times 9 =$ ___ $9 \times 7 =$ ___ $3 \times 6 =$ ___
   $32 \div 4 = 8$ ___ ÷ ___ = ___   ___ ÷ ___ = ___   ___ ÷ ___ = ___
   $32 \div 8 = 4$ ___ ÷ ___ = ___   ___ ÷ ___ = ___   ___ ÷ ___ = ___

b. $7 \times 6 =$ ___   d. $4 \times 6 =$ ___   f. $5 \times 7 =$ ___   h. $8 \times 7 =$ ___
   $6 \times 7 =$ ___ $6 \times 4 =$ ___ $7 \times 5 =$ ___ $7 \times 8 =$ ___
   ___ ÷ ___ = ___   ___ ÷ ___ = ___   ___ ÷ ___ = ___   ___ ÷ ___ = ___
   ___ ÷ ___ = ___   ___ ÷ ___ = ___   ___ ÷ ___ = ___   ___ ÷ ___ = ___

> You just reviewed how division is related to multiplication.

Applying Division

WHAT DO YOU KNOW ABOUT THE OPERATION OF DIVISION?

When you divide, you separate a number into equal groups.
Look at the problem $6 \div 3 = 2$.
The problem asks you to separate 6 into 3 groups of 2: 6 = ② ② ②
Division is repeated subtraction: $6 - 2 - 2 - 2 = 0$.
You subtract 3 groups of 2 from 6 to get 0.
This can also be expressed as separating 6 into 3 groups of 2.
The division problem is $6 \div 3 = 2$.

▶ Write each repeated subtraction problem as a division problem. The first two problems have been completed for you.

a. $10 - 2 - 2 - 2 - 2 - 2 = 0$
The division problem is $10 \div 5 = 2$.

b. $10 - 5 - 5 = 0$
The division problem is $10 \div 2 = 5$.

c. $30 - 5 - 5 - 5 - 5 - 5 - 5 = 0$
The division problem is ___ ÷ ___ = ___.

d. $42 - 7 - 7 - 7 - 7 - 7 - 7 = 0$
The division problem is ___ ÷ ___ = ___.

e. $27 - 9 - 9 - 9 = 0$
The division problem is ___ ÷ ___ = ___.

f. $32 - 8 - 8 - 8 - 8 = 0$
The division problem is ___ ÷ ___ = ___.

g. $24 - 6 - 6 - 6 - 6 = 0$
The division problem is ___ ÷ ___ = ___.

h. $25 - 5 - 5 - 5 - 5 - 5 = 0$
The division problem is ___ ÷ ___ = ___.

> You just reviewed how to use repeated subtraction in division.

> Together, write four division problems with a two-digit divisor, a two-digit dividend, and a quotient that has no remainder. Use repeated subtraction to illustrate each problem. Discuss the problems.

Applying Division

PART TWO: Learn About Division

Study the problem that Karina's teacher wrote on the chalkboard. As you study, think about the two ways that Karina solved the problem.

Problem: There are 90 crayons in all. The crayons fill 10 boxes evenly. How many crayons are in each box?

Two Ways That Karina Solved the Problem	
1. I draw a picture and count how many in each group.	(picture of 10 boxes with 9 dots each)
2. I use fact families. I write a multiplication sentence. I rewrite it as a division sentence.	$10 \times \underline{?} = 90 \rightarrow 10 \times \underline{9} = 90$ $90 \div 10 = 9$

Answer: There are 9 crayons in each box.

When you divide, you separate a number into equal amounts.

A division problem has three parts. Divide the dividend by the divisor. The answer is the quotient.

$$48 \div 6 = 8$$
dividend   divisor   quotient

You can write a division problem two ways:

Division Sentence   **Long Division**
$12 \div 4 = 3$

divisor — $4\overline{)12}$ — quotient / dividend

You use **division** to find a quotient.

▶ To divide, separate a number into equal amounts.

▶ Use multiplication facts to help find a quotient.

60 Applying Division

Study another one of Karina's problems. Think about the two ways that Karina solved the problem. Then do Numbers 1 through 4.

Problem: There are 25 markers. If 5 students are to divide the markers equally, how many markers will each student have?

1. Separate 25 into 5 equal groups. Count how many in each group.

2. Use multiplication facts to write a multiplication sentence. $5 \times ? = 25 \rightarrow 5 \times \underline{5} = 25$

   Rewrite it as a division sentence. $25 \div 5 = 5$

Answer: Each student will have 5 markers.

1. Karina counted 27 drawings of dogs. She knew that each student in her group had made 3 drawings. How many students are in Karina's group?
   - Ⓐ 6 students
   - Ⓑ 9 students
   - Ⓒ 7 students
   - Ⓓ 8 students

2. Karina knew that 8 students drew pictures of cats. Each student drew the same number of cat pictures. Karina counted 24 pictures of cats in all. How many pictures of cats did each student draw?
   - Ⓐ 3 pictures
   - Ⓑ 8 pictures
   - Ⓒ 5 pictures
   - Ⓓ 7 pictures

3. Today, 6 students in Karina's class made 24 watercolor paintings. If each student painted the same number of watercolors, how many paintings did each student do?
   - Ⓐ 5 paintings
   - Ⓑ 6 paintings
   - Ⓒ 4 paintings
   - Ⓓ 3 paintings

4. Karina painted pictures of the 4 seasons. She painted the same number of pictures of each season, for a total of 28 pictures. How many pictures did she paint of each season?
   - Ⓐ 7 pictures
   - Ⓑ 9 pictures
   - Ⓒ 5 pictures
   - Ⓓ 8 pictures

Talk about your answers to questions 1–4. Tell why you chose the answers you did.

Applying Division

PART THREE: Check Your Understanding

Remember: You use division to find a quotient.

▶ To divide, separate a number into equal amounts.

▶ Use multiplication facts to help find a quotient.

Solve this problem. As you work, ask yourself, "How do I separate this number into equal amounts?"

5. There were 36 students in Karina's art class. The art teacher divided them into groups of 6 students. How many groups did the art teacher form?
   - Ⓐ 9 groups
   - Ⓑ 7 groups
   - Ⓒ 4 groups
   - Ⓓ 6 groups

Solve another problem. As you work, ask yourself, "How can I use multiplication facts to help me find the quotient?"

6. Each of the 6 students in Karina's group made the same number of drawings. At the end of class, the teacher collected 54 drawings from Karina's group. How many drawings did each student in Karina's group make?
   - Ⓐ 6 drawings
   - Ⓑ 9 drawings
   - Ⓒ 8 drawings
   - Ⓓ 7 drawings

Applying Division

Look at the answer choices for each question.
Read why each answer choice is correct or not correct.

5. There were 36 students in Karina's art class. The art teacher divided them into groups of 6 students. How many groups did the art teacher form?

   Ⓐ 9 groups

   This answer is not correct because 36 divided into 6 equal amounts is 6, not 9.

   Ⓑ 7 groups

   This answer is not correct because 36 divided into 6 equal amounts is 6, not 7.

   Ⓒ 4 groups

   This answer is not correct because 36 divided into 6 equal amounts is 6, not 4.

   ● 6 groups

   This answer is correct because 36 divided into 6 equal amounts is 6.

6. Each of the 6 students in Karina's group made the same number of drawings. At the end of class, the teacher collected 54 drawings from Karina's group. How many drawings did each student in Karina's group make?

   Ⓐ 6 drawings

   This answer is not correct because $6 \times 6 = 36$, which you can rewrite as $36 \div 6 = 6$. Multiplication facts tell you that $6 \times 6 = 36$, not 54.

   ● 9 drawings

   This answer is correct because $6 \times 9 = 54$, which you can rewrite as $54 \div 6 = 9$.

   Ⓒ 8 drawings

   This answer is not correct because $6 \times 8 = 48$, which you can rewrite as $48 \div 6 = 8$. Multiplication facts tell you that $6 \times 8 = 48$, not 54.

   Ⓓ 7 drawings

   This answer is not correct because $6 \times 7 = 42$, which you can rewrite as $42 \div 6 = 7$. Multiplication facts tell you that $6 \times 7 = 42$, not 54.

Applying Division

PART FOUR: Learn More About Division

You use long division to divide a two-digit dividend by a one-digit divisor.

▶ Divide the tens. Then divide the ones. Check your answer by multiplying the quotient by the divisor.

▶ Line up the tens and ones in the quotient with the tens and ones in the dividend.

Problem: $56 \div 4 = ?$

| 1. Divide the tens: 4 tens goes into 5 tens 1 time.
Multiply.
$4 \times 1 = 4$
Subtract.
$5 - 4 = 1$

1
$4\overline{)56}$
$\underline{-4}$
1 | 2. Bring down the ones digit.

1
$4\overline{)56}$
$\underline{-4}$
16 | 3. Divide the ones: 4 goes into 16 4 times.
Multiply.
$4 \times 4 = 16$
Subtract.
$16 - 16 = 0$

14
$4\overline{)56}$
$\underline{-4}$
16
$\underline{-16}$
0 | 4. Check your quotient or answer.

14
$\underline{\times4}$
56 |

**Karina learned how to draw flowers in art class.
Do Numbers 7 through 10.**

7. Karina drew 2 pictures of each flower. In all, she drew 26 pictures. How many different flowers did she draw?

   Ⓐ 13 flowers   Ⓒ 9 flowers
   Ⓑ 10 flowers   Ⓓ 12 flowers

8. Karina's art teacher told the students they would each need 3 brushes to paint roses. There were 42 brushes available. How many students could paint roses at one time?

   Ⓐ 10 students   Ⓒ 16 students
   Ⓑ 14 students   Ⓓ 13 students

9. Some students drew daisies. Each daisy had 10 petals. Karina counted 70 petals in all. How many daisies did Karina see?

   Ⓐ 10 daisies   Ⓒ 12 daisies
   Ⓑ 8 daisies   Ⓓ 7 daisies

10. Karina's art teacher put the flower drawings into folders. She put 5 drawings into each folder. If there were 75 drawings, how many folders did she need?

    Ⓐ 17 folders   Ⓒ 15 folders
    Ⓑ 11 folders   Ⓓ 13 folders

Applying Division

Karina read an art book. Read this part of one chapter about paintings. Then do Numbers 11 through 14.

Painting the World Around You

Three types of paintings are still life, cityscape, and landscape. A still-life painting shows objects such as a bowl of fruit or a pile of books. A cityscape painting can show buildings, cars, people, and other things in a city. A landscape painting shows things in nature, like an ocean, a sunset, or a forest. Landscape paintings can also show living things, like animals and plants.

11. Karina made a painting of a building on her street. The building had 6 floors. Each floor had the same number of windows. Karina counted 72 windows in her painting. How many windows were on each floor?

    Ⓐ 15 windows
    Ⓑ 12 windows
    Ⓒ 10 windows
    Ⓓ 14 windows

12. Karina painted 4 landscapes at each place she visited on her vacation. When she got home, she had 56 paintings. How many places did she visit?

    Ⓐ 16 places   Ⓒ 11 places
    Ⓑ 12 places   Ⓓ 14 places

13. Karina wanted to try some still-life paintings. She had 5 bowls and 65 apples. She placed the same number of apples into each bowl. How many apples did she place into each bowl?

    Ⓐ 13 apples   Ⓒ 12 apples
    Ⓑ 9 apples   Ⓓ 10 apples

14. Karina keeps her paintings in boxes, one box for each type. She has 51 paintings, and each box has the same number of paintings. How many paintings are in each box?

    Ⓐ 14 paintings
    Ⓑ 17 paintings
    Ⓒ 23 paintings
    Ⓓ 15 paintings

Applying Division

PART FIVE: Prepare for a Test

▶ A test question about division may ask for a quotient.

▶ A test question about division may ask you to do long division to find the quotient of numbers with more than one digit.

Karina wanted to try some other art forms. She chose to work with clay. Read about art that uses clay. Then do Numbers 15 and 16.

From Clay to Pots

Clay can be used to make many kinds of art. The two most popular uses for clay are pottery and sculpture. For pottery, artists use clay to make things such as bowls, pots, plates, cups, and vases. They put wet clay on a spinning disk called a *potter's wheel*. While the clay spins, potters press on the clay with their hands to form a round shape. They then stretch and trim the clay until it is the right shape. Next, they put the pot into a special oven, called a *kiln*, where it is baked until it is hard. Potters either leave the pot plain or paint it.

Applying Division

15. Karina took a pottery class. The teacher gave each student 36 ounces of clay. It takes 4 ounces of clay to make 1 small bowl. How many small bowls can each student make?
    - Ⓐ 7 bowls
    - Ⓑ 14 bowls
    - Ⓒ 9 bowls
    - Ⓓ 6 bowls

Applying Division

16. The students in Karina's class made a total of 40 large bowls. The kiln at the art school can hold 10 large bowls at a time. How many times will they have to fill the kiln to bake all of the bowls?
    - Ⓐ 6 times
    - Ⓑ 4 times
    - Ⓒ 8 times
    - Ⓓ 5 times

Read more about how Karina used clay to complete an art project. Then do Numbers 17 and 18.

From Clay to Sculpture

To make a sculpture from clay, potters model the clay by hand or with special tools. The artists can shape the clay into almost any form they can imagine. Often, sculptures are figures of people or animals, but they can also be of other things, like trees or cars. After the potters model the clay, they either bake the sculpture or let it dry until it is hard. Some artists then paint their sculptures.

Applying Division

17. Karina liked working with clay so much that she made a sculpture that she called *Friendship*. The sculpture was a circle of hands of different sizes and shapes. When she was done, she had formed 95 fingers and thumbs. How many hands were in Karina's sculpture?
    - Ⓐ 17 hands
    - Ⓑ 21 hands
    - Ⓒ 13 hands
    - Ⓓ 19 hands

Applying Division

18. When she was finished, Karina found that she had spent a total of 57 hours on her bowls and sculpture. She wanted to find how many weeks it had taken her to make all the clay forms. She knew that she had worked 3 hours a week on them. How many weeks did Karina spend working on her clay forms?
    - Ⓐ 19 weeks
    - Ⓑ 27 weeks
    - Ⓒ 21 weeks
    - Ⓓ 29 weeks

Applying Division

PART ONE: Read a Story

Read the story about Jaina and a summer surprise.
Then do Numbers 1 through 6.

Jaina was bored. "There's never anything to do around here in the summer," she said to herself as she sat on her front porch.

Soon, a long line of big trucks drove up Jaina's road. On the side of every truck, painted in huge red letters, was the word CARNIVAL. Excited, Jaina jumped on her bike and followed the trucks to the field at the edge of town.

Some of the trucks parked along the edges of the field. Other trucks parked in 6 rows across the field. Jaina watched in delight as carnival workers began unloading the trucks.

First, the workers lifted up the sides of the smaller trucks that were parked in the center of the field. Inside were games and food stands. The games were already set up inside the trucks. Even the prizes were hanging from the ceilings and walls!

Then the workers set up the carnival rides. They had been folded up on bigger trucks. Big piles of metal pieces turned into a giant Ferris wheel, Tilt-o-Whirl, Caterpillar, carousel, and other rides!

Next, the carnival workers opened the trucks that were parked along the sides of the field. Jaina had wondered why these trucks had large open windows. Now she knew. One by one, the workers led groups of animals from the trucks. First came 3 sheep. Then came 4 goats. There was even a camel! The workers put up fences and set out large buckets of water and food for the animals. Soon, the children's petting zoo was complete.

Finally, one of the carnival workers led 2 black-and-white-spotted ponies from the last truck. The woman led the ponies to a fenced-in area. Jaina read the sign on the fence gate. It said, Pony Rides— 2 tickets.

Jaina raced home on her bike. "Maybe summer isn't so boring after all," she thought.

▶ Applying Subtraction   ▶ Applying Multiplication   ▶ Applying Division

Applying Subtraction

1. In all, Jaina counted 50 trucks. There were 18 trucks carrying the games. How many other trucks were there?
   - Ⓐ 22 trucks
   - Ⓑ 28 trucks
   - Ⓒ 32 trucks
   - Ⓓ 38 trucks

Applying Multiplication

4. The Ferris wheel had 26 cars. Each car held 3 people. How many people could ride the Ferris wheel at one time?
   - Ⓐ 74 people
   - Ⓑ 86 people
   - Ⓒ 68 people
   - Ⓓ 78 people

Applying Subtraction

2. On Sunday, 620 people went to the carnival. On Monday, 535 people went to the carnival. How many more people went on Sunday than on Monday?
   - Ⓐ 85 people
   - Ⓑ 125 people
   - Ⓒ 115 people
   - Ⓓ 95 people

Applying Division

5. Jaina counted the trucks parked in the 6 rows. The same number of trucks were in each row. There were 24 trucks in all. How many trucks were in each row?
   - Ⓐ 9 trucks
   - Ⓑ 4 trucks
   - Ⓒ 8 trucks
   - Ⓓ 30 trucks

Applying Multiplication

3. Jaina rode the carousel 4 times. Each ride cost 5 tickets. How many tickets did Jaina use for the carousel?
   - Ⓐ 9 tickets
   - Ⓑ 18 tickets
   - Ⓒ 20 tickets
   - Ⓓ 24 tickets

Applying Division

6. The pony ride costs 2 tickets. On the first day, 68 tickets were collected at the pony ride. How many people took a pony ride?
   - Ⓐ 34 people
   - Ⓑ 32 people
   - Ⓒ 39 people
   - Ⓓ 136 people

Strategies 4–6 Review

PART TWO: Read an Article

Read the article about alligators and crocodiles. Then do Numbers 7 through 12.

How are alligators and crocodiles alike?

Alligators and crocodiles are reptiles. They have many things in common. They both have a long snout, four legs, and a powerful tail. Both swim near the top of the water. You can often see only their eyes and nostrils above the water. They both use their tail to swim. They both can close their nostrils. They can even close their ears!

Both alligators and crocodiles live in hot areas, in marshes or near rivers. However, Florida is the only state where both alligators and crocodiles live in the wild.

Another way that alligators and crocodiles are alike is how they hatch their young. The mothers build a nest out of plants or mud. Then they lay their eggs in the nest. The sun warms the eggs until they hatch. Young alligators or crocodiles call to their mother. When their mother hears the calls, she carries the babies in her mouth to the water.

Both alligators and crocodiles eat fish, frogs, birds, and other small animals. You can be safe from alligators and crocodiles as long as you don't bother or tease them.

What are some differences between alligators and crocodiles?

The biggest difference between alligators and crocodiles is the shape of their snout. The fourth tooth on each side of a crocodile's lower jaw is long. The crocodile has notches in the sides of its snout that show these teeth. The alligator has a wider snout.

Crocodiles have bony plates on their belly. Alligators do not have these plates.

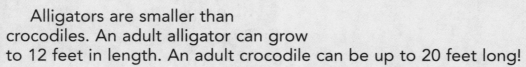

Alligators are smaller than crocodiles. An adult alligator can grow to 12 feet in length. An adult crocodile can be up to 20 feet long!

Applying Subtraction

7. Millions of years ago, crocodiles grew to 40 feet in length. How much longer were they than today's longest crocodiles?
   - Ⓐ 30 feet
   - Ⓑ 20 feet
   - Ⓒ 10 feet
   - Ⓓ 18 feet

Applying Multiplication

10. An alligator usually lays about 32 eggs in a nest. About how many eggs would be in 3 alligator nests?
    - Ⓐ 93 eggs
    - Ⓑ 96 eggs
    - Ⓒ 92 eggs
    - Ⓓ 64 eggs

Applying Subtraction

8. Most Chinese alligators are 5 feet long. The largest known American alligator was 19 feet long. How much longer was the American alligator than the Chinese alligator?
   - Ⓐ 17 feet
   - Ⓑ 13 feet
   - Ⓒ 15 feet
   - Ⓓ 14 feet

Applying Division

11. Crocodile eggs hatch in about 91 days. How many weeks is this?
    - Ⓐ 17 weeks
    - Ⓑ 13 weeks
    - Ⓒ 21 weeks
    - Ⓓ 11 weeks

Applying Multiplication

9. In total, how many legs do 18 alligators have?
   - Ⓐ 72 legs
   - Ⓑ 42 legs
   - Ⓒ 64 legs
   - Ⓓ 86 legs

Applying Division

12. Each alligator measures the same length. How long is each alligator?

    - Ⓐ 9 feet
    - Ⓑ 7 feet
    - Ⓒ 10 feet
    - Ⓓ 6 feet

Strategy Seven: CONVERTING TIME AND MONEY

PART ONE: Think About Time and Money

WHAT DO YOU KNOW ABOUT CONVERTING TIME?

A clock shows time in minutes and hours.

The short hand on a clock is the *hour hand*.

The long hand on a clock is the *minute hand*.

There are 60 minutes in 1 hour.

Use A.M. to show morning time and P.M. to show afternoon and evening time.

▶ Convert each group of minutes to hours and minutes.

a. 65 minutes = _____ hour and _____ minutes
b. 82 minutes = _____ hour and _____ minutes
c. 94 minutes = _____ hour and _____ minutes
d. 106 minutes = _____ hour and _____ minutes
e. 114 minutes = _____ hour and _____ minutes
f. 122 minutes = _____ hours and _____ minutes

▶ Add the hour hand and the minute hand to show each time on the clock.

a.
12:35 A.M. or P.M.

c.
3:58 A.M. or P.M.

e.
5:45 A.M. or P.M.

b.
10:22 A.M. or P.M.

d.
11:39 A.M. or P.M.

f.
7:06 A.M. or P.M.

> You just reviewed how minutes can be converted into hours and minutes and how a clock shows time.

WHAT DO YOU KNOW ABOUT CONVERTING MONEY?

A quarter has a value of 25¢.	A nickel has a value of 5¢.
A dime has a value of 10¢.	A penny has a value of 1¢.

▶ Convert each amount into coins.
   a. What three coins have a value of 27¢?
   _____
   b. What three coins have a value of 36¢?
   _____
   c. What three coins have a value of 51¢?
   _____
   d. What four coins have a value of 26¢?
   _____
   e. What four coins have a value of 76¢?
   _____
   f. What four coins have a value of 80¢?
   _____

▶ What is the value of each group of coins?
   a. Three quarters, one dime, and one nickel = _____.
   b. Two quarters, one dime, one nickel, and two pennies = _____.
   c. Three quarters, one dime, one nickel, and four pennies = _____.
   d. Three dimes, three nickels, and three pennies = _____.

> You just reviewed information about the total value of coin combinations.

Separately, write three time problems and three coin problems. When finished, ask your partner to solve your problems, while you solve your partner's problems.

Converting Time and Money   73

PART TWO: Learn About Time

Study the clocks and read what Leon learned about clocks and time. As you study, think about how to use a clock to tell time.

An analog clock has two hands.
- The shorter hand is the hour hand. The hour hand takes 60 minutes to move from one number to the next.
- The longer hand is the minute hand. The minute hand takes 5 minutes to move from one number to the next.

A digital clock shows time in hours and minutes.
- Both clocks show 6:10, or 10 minutes after 6.

You may want to know how much time has gone by.
- Leon found that 3 hours and 30 minutes had passed since breakfast. He counted by 1s to find that 3 hours had passed. He skip-counted by 5s to find that 30 more minutes had passed.

You may want to add hours and minutes to one time to find another time. You can count on, or you can use addition. Remember to change each 60 minutes to 1 hour. Be sure to add hours and minutes separately.

You use a clock to tell **time** and to tell how much time has gone by.

▶ Look at the shorter hand on the clock to tell the hour. Look at the longer hand to tell how many minutes there are after or before an hour.

▶ Count by 1s to find how many hours have gone by. Skip-count by 5s to find how many minutes have gone by.

▶ To add times, change each 60 minutes to 1 hour. Add all the hours. Then add all the minutes.

▶ Use A.M. to show morning time. Use P.M. to show afternoon and evening time.

Converting Time and Money

Study how Leon solved this problem. Then do Numbers 1 through 4.

Problem: A girl starts her homework at the time shown on the clock. She finishes her homework 70 minutes later. What time does she finish her homework?

I know that 60 minutes = 1 hour. So, 70 minutes = 1 hour and 10 minutes.
- In my mind, I picture 3:27 on a clock. I move the hour hand ahead 1 hour. I move the minute hand ahead 10 minutes. The clock now says 4:37.
- I can also add 1 hour and 10 minutes to 3:27. I line up minutes and hours. I add minutes. Then I add hours.

Answer: The girl finishes her homework at 4:37 P.M.

1. Leon gets up at the time shown. What time does he get up?
   - Ⓐ 7:05 A.M.
   - Ⓑ 6:55 A.M.
   - Ⓒ 8:15 A.M.
   - Ⓓ 4:10 A.M.

2. Use the clock in Number 1. Leon leaves at 8:25 A.M. How long does it take Leon to leave his house?
   - Ⓐ 4 hours
   - Ⓑ 1 hour, 15 minutes
   - Ⓒ 2 hours, 25 minutes
   - Ⓓ 1 hour, 20 minutes

3. School ends at the time shown. On Friday, Leon stayed 90 minutes later. What time did Leon leave school?
   - Ⓐ 3:45 P.M.
   - Ⓑ 4:10 P.M.
   - Ⓒ 4:45 P.M.
   - Ⓓ 3:55 P.M.

4. Leon raked leaves for 80 minutes. He started at the time shown on the clock. What time did he finish?
   - Ⓐ 12:08 P.M.
   - Ⓑ 11:58 A.M.
   - Ⓒ 10:58 A.M.
   - Ⓓ 11:28 A.M.

Talk about your answers to questions 1–4. Tell why you chose the answers you did.

Converting Time and Money

PART THREE: Check Your Understanding

Remember: You use a clock to tell time and to tell how much time has gone by.

▶ Look at the shorter hand on the clock to tell the hour. Look at the longer hand to tell how many minutes there are after or before an hour.

▶ Count by 1s to find how many hours have gone by. Skip count by 5s to find how many minutes have gone by.

▶ To add times, change each 60 minutes to 1 hour. Add all the hours. Then add all the minutes.

▶ Use A.M. to show morning time. Use P.M. to show afternoon and evening time.

Solve this problem. As you work, ask yourself, "How can counting by 1s and 5s help me know how much time has gone by?"

5. Leon and other members of the school band marched in a parade. The parade started at the time shown on the first clock. It ended at the time shown on the second clock. How long did the parade last?

Ⓐ 2 hours, 15 minutes
Ⓑ 1 hour, 15 minutes
Ⓒ 1 hour, 45 minutes
Ⓓ 2 hours

Solve another problem. As you work, ask yourself, "How many minutes are in 1 hour? How can I count on or use addition to find the time?"

6. After the parade, the band played at a town picnic. The concert started at the time shown on the clock. It lasted for 75 minutes. What time did the concert end?

Ⓐ 5:49 P.M.
Ⓑ 5:19 P.M.
Ⓒ 4:49 P.M.
Ⓓ 5:54 P.M.

**Look at the answer choices for each question.
Read why each answer choice is correct or not correct.**

5. Leon and other members of the school band marched in a parade. The parade started at the time shown on the first clock. It ended at the time shown on the second clock. How long did the parade last?

   Ⓐ 2 hours, 15 minutes

   This answer is not correct because it is 30 minutes more than the correct amount of time. If you count hours by 1s and minutes by 5s, you get 1 hour, 45 minutes.

   Ⓑ 1 hour, 15 minutes

   This answer is not correct because it is 30 minutes less than the correct amount of time. If you count hours by 1s and minutes by 5s, you get 1 hour, 45 minutes.

   ● 1 hour, 45 minutes

   This answer is correct because if you count hours by 1s and minutes by 5s, you get 1 hour, 45 minutes.

   Ⓓ 2 hours

   This answer is not correct because it is 15 minutes more than the correct amount of time. If you count hours by 1s and minutes by 5s, you get 1 hour, 45 minutes.

6. After the parade, the band played at a town picnic. The concert started at the time shown on the clock. It lasted for 75 minutes. What time did the concert end?

   ● 5:49 P.M.

   This answer is correct because 75 minutes is 1 hour and 15 minutes. When you count on from 4:34 or add 75 minutes to 4:34, you get 5:49.

   Ⓑ 5:19 P.M.

   This answer is not correct because it is 30 minutes earlier than the correct time. You may not have changed 75 minutes into 1 hour and 15 minutes. When you add 75 minutes to 4:34, you get 5:49, not 5:19.

   Ⓒ 4:49 P.M.

   This answer is not correct because it is 1 hour earlier than the correct time. You may not have added hours or counted on correctly. When you add 75 minutes to 4:34, you get 5:49, not 4:49.

   Ⓓ 5:54 P.M.

   This answer is not correct because it is 5 minutes later than the correct time. You may not have added minutes or counted on correctly. When you add 75 minutes to 4:34, you get 5:49, not 5:54.

Converting Time and Money

PART FOUR: Learn More About Money

You use what you know about the value of coins and bills to count **money**.

▶ To find the total value of coins, count all the coins. Start with the coins that have the highest value.

1 quarter = 25¢	1 dime = 10¢	1 nickel = 5¢	1 penny = 1¢

Count by 25s. 25, 50,	Count by 10s. 60, 70, 80, 90,	Count by 5s. 95, 100, 105,	Count by 1s. 106, 107

The total value of the coins shown on the chart is $1.07.

▶ To find the total value of a group of bills, or paper money count each group of bills. Start with the group of bills having the highest value.

Leon raised money for the school band at a used-book sale. Do Numbers 7 through 10.

7. Leon sold a used copy of the book *Stuart Little* for the coins shown. How much did the book cost?

   Ⓐ $1.22   Ⓒ $1.32
   Ⓑ $1.37   Ⓓ $1.27

8. A book of nursery rhymes sold for 57¢. Which coins total 57¢?

   Ⓐ one quarter, one dime, and seven pennies
   Ⓑ two quarters, one nickel, and two pennies
   Ⓒ four dimes, two nickels, and two pennies
   Ⓓ one quarter, one dime, one nickel, and seven pennies

9. After two hours of sales, Leon counted the bills in the money box. He counted $22.00 in bills. Which bills total $22.00?

   Ⓐ one ten, one five, and two ones
   Ⓑ three fives and three ones
   Ⓒ two tens and two ones
   Ⓓ two tens, one five, and seven ones

10. Leon's friend Ann paid for a book by Beverly Cleary and got 29¢ in change. Which group of coins totals 29¢?

    Ⓐ one dime, one nickel, and four pennies
    Ⓑ two dimes and four pennies
    Ⓒ two dimes, two nickels, and four pennies
    Ⓓ one dime, three nickels, and four pennies

Read how Leon helps out at his mother's bookstore.
Then do Numbers 11 through 14.

On Saturday mornings, Leon walks to his mom's bookstore. Leon helps his mom unpack boxes of new books and put them on shelves. He knows to put fiction books in alphabetical order, by the author's last name. Nonfiction books are grouped by their topics. Each week, Leon's mom picks some books to put on sale. Leon helps put sale stickers on those books.

11. On the way to the bookstore, Leon stopped to buy muffins. He paid with the coins shown. How much did he pay for the muffins?

Ⓐ $0.93
Ⓑ $1.03
Ⓒ $0.98
Ⓓ $1.08

12. Leon put a sale sticker of $17.00 on a new children's dictionary. The man who bought the dictionary paid the exact amount. Which of the sets of bills did the man use?

Ⓐ two tens and seven ones
Ⓑ one ten, one five, and two ones
Ⓒ one ten, one five, and three ones
Ⓓ one ten, two fives, and four ones

13. A boy bought a paperback book about bugs. He paid the exact amount with 2 quarters, 5 dimes, 2 nickels, and 4 pennies. What was the cost of the book?

Ⓐ $0.99
Ⓑ $1.04
Ⓒ $1.14
Ⓓ $1.19

14. Leon counted out 86¢ in change for one customer. Which group of coins could Leon have given the customer?

Ⓐ three quarters, one dime, and one penny
Ⓑ two quarters, two dimes, one nickel, and one penny
Ⓒ five dimes, four nickels, and one penny
Ⓓ three quarters, three nickels, and one penny

Converting Time and Money

PART FIVE: Prepare for a Test

▶ A test question about time may ask for the time shown on a clock or how many hours and minutes have gone by.

▶ A test question about time may ask you to change minutes to hours or to add minutes and hours.

▶ A test question about money may ask for the total value of coins or bills.

The school drama club put on a musical play. Read how Leon took part in the play. Then do Numbers 15 and 16.

Leon Plays for the Play

The drama club decided to put on the musical play *Sleeping Beauty*. Leon's band teacher asked Leon if he would like to play the trumpet in the band. Leon said, "Yes." The band and the actors practiced for two months. Finally, they were ready to perform *Sleeping Beauty* for an audience.

Converting Time and Money

15. The play began at the time shown on the first clock. It ended at the time shown on the second clock. How long was the play?

- Ⓐ 2 hours, 15 minutes
- Ⓑ 3 hours, 25 minutes
- Ⓒ 2 hours, 35 minutes
- Ⓓ 2 hours

Converting Time and Money

16. Each member of the band had to buy a purple ribbon sash to wear for the performance. Study the coins that Leon gave his teacher to pay for his sash. What was the cost of the sash?

- Ⓐ $1.45
- Ⓑ $1.50
- Ⓒ $1.05
- Ⓓ $1.40

Converting Time and Money

Read what Leon learned about a famous trumpet player named Louis Armstrong. Then do Numbers 17 and 18.

Louis Armstrong

Leon read a book about Louis Armstrong, a jazz musician. Leon learned that Armstrong grew up in a poor neighborhood in New Orleans, Louisiana, in the early 1900's. When Louis was 12, he learned to play the cornet, a kind of trumpet. In time, he became a world-famous trumpet player. People sometimes call Louis Armstrong by his nickname, Satchmo. Louis Armstrong died in 1971.

Converting Time and Money

17. Leon started reading the book about Louis Armstrong at the time shown on the clock. He finished the book 85 minutes later. What time did Leon finish reading the book?

    Ⓐ 2:39 P.M.
    Ⓑ 3:19 P.M.
    Ⓒ 1:39 P.M.
    Ⓓ 2:09 P.M.

Converting Time and Money

18. At the music store, Leon's mom bought a CD of Louis Armstrong's greatest hits. Leon's mom paid exactly $26.00 for the CD. Which bills did she use?

    Ⓐ one ten, one five, and ten ones
    Ⓑ two tens, one five, and six ones
    Ⓒ one ten, one five, and four ones
    Ⓓ two tens, one five, and one one

Strategy Eight: CONVERTING CUSTOMARY AND METRIC MEASURES

PART ONE: Think About Customary and Metric Measures

WHAT DO YOU KNOW ABOUT CUSTOMARY MEASURES?

To measure length, width, and height, use units like inch, foot, or yard.
To measure capacity, or "how much" of a liquid or solid, use units like ounce, cup, pint, quart, or gallon.
To measure weight, or how heavy something is, use units like ounce or pound.
Customary measures are used chiefly in the United States.

▶ Answer each question about the customary measures for length, width, and height.

  a. How many inches are in one foot? _____
  b. How many feet are in one yard? _____
  c. How many inches are in one yard? _____

▶ Answer each question about the customary measures for capacity.

  a. How many ounces are in one cup? _____
  b. How many cups are in one pint? _____
  c. How many pints are in one quart? _____
  d. How many quarts are in one gallon? _____

▶ Answer each question about the customary measures for weight.

  a. How many ounces are in one pound? _____
  b. How many pounds are in one ton? _____

▶ Circle the customary measure that is the larger of the two measures.

  a. 25 inches or 2 feet   d. 6 quarts or 2 gallons
  b. 18 ounces or 1 pound   e. 40 inches or 1 yard
  c. 5 pints or 2 quarts   f. 5 feet or 2 yards

You just reviewed how customary measures can be converted from one unit to another.

WHAT DO YOU KNOW ABOUT METRIC MEASURES?

To measure length, width, and height, use units like centimeters and meters.
To measure capacity, use a unit like liters.
To measure weight, use units like grams and kilograms.
Metric measures were developed in France in 1790.
Today the metric system is used in Canada, Great Britain, and several other European countries.

▶ Answer each question about the metric measures for length, width, and height.
   a. How many centimeters are in one meter? _____
   b. How many meters are in one thousand centimeters? _____

▶ Answer the question about a metric measure for capacity.
   a. What is the metric measure most often used to measure a liquid? _____

▶ Answer each question about the metric measures for weight.
   a. How many kilograms are in one thousand grams? _____
   b. How many grams are in one kilogram? _____

▶ Circle the metric measure that is the larger of the two measures.
   a. 100 grams or 1 kilogram
   b. 1.5 liters or $1\frac{2}{3}$ liters
   c. 90 centimeters or 1 meter
   d. 1 meter or 1 kilometer
   e. 1 meter or 1 centimeter

You just reviewed how metric measures can be converted from one unit to another.

Separately, write three problems based on customary measures and three problems based on metric measures. When finished, ask your partner to solve your problems, while you solve your partner's problems.

Converting Customary and Metric Measures

PART TWO: Learn About Customary and Metric Measures

Study the measurement chart that Lena's teacher made. As you study, think about the different ways that objects can be measured.

UNITS OF MEASUREMENT	
Customary	Metric
inch, foot, yard, mile	millimeter, centimeter, meter, kilometer
ounce, cup, pint, quart, gallon	liter
ounce, pound, ton	milligram, gram, kilogram

You use different units of measurement to measure different things. To measure length, height, or width, use units like inches, feet, or meters. To measure capacity, or "how much" of a liquid or a solid, use units like cups, pints, or liters. To measure weight, or how heavy something is, use units like pounds or grams.

If measurements are given in different units, change them to the same unit. To change a measurement to a smaller unit, multiply. To change a measurement to a larger unit, divide.

Multiply the number of	to get the number of
feet by 12	inches
yards by 3	feet
meters by 100	centimeters
pints by 2	cups
quarts by 32	ounces
gallons by 4	quarts
pounds by 16	ounces

Divide the number of	to get the number of
inches by 12	feet
feet by 3	yards
centimeters by 100	meters
cups by 2	pints
ounces by 32	quarts
quarts by 4	gallons
ounces by 16	pounds

You use **measurement** to find the size of something.

▶ Use multiplication to change a measurement from a larger unit to a smaller unit.
▶ Use division to change a measurement from a smaller unit to a larger unit.

Converting Customary and Metric Measures

Lena helped her dad clean up the cellar. For fun, she determined the measure of some things. Her list is below. Study the measurements she made. Think about ways to change measurements from one unit to another. Then do Numbers 1 through 4.

ITEM	Nail	Garden Hose	Paint	Workbench
MEASUREMENT	3 inches	15 feet	10 gallons	3 meters (length)

1. Lena wanted to change the measurement of the garden hose to yards. How many yards long is the garden hose?
   Ⓐ 45 yards
   Ⓑ 12 yards
   Ⓒ 18 yards
   Ⓓ 5 yards

2. Lena measured 10 gallons of paint. How many quarts of paint is that?
   Ⓐ 40 quarts
   Ⓑ 12 quarts
   Ⓒ 5 quarts
   Ⓓ 30 quarts

3. If Lena lined up 8 nails end to end, they would measure 24 inches. How many feet is 24 inches?
   Ⓐ 8 feet
   Ⓑ 2 feet
   Ⓒ 4 feet
   Ⓓ 12 feet

4. Lena wrote the measurement of the workbench in centimeters. What is the length of the workbench in centimeters?
   Ⓐ 330 centimeters
   Ⓑ 30 centimeters
   Ⓒ 300 centimeters
   Ⓓ 130 centimeters

Talk about your answers to questions 1–4. Tell why you chose the answers you did.

Converting Customary and Metric Measures

PART THREE: Check Your Understanding

Remember: You use measurement to find the size of something.

▶ Use multiplication to change a measurement from a larger unit to a smaller unit.

▶ Use division to change a measurement from a smaller unit to a larger unit.

Solve this problem. As you work, ask yourself, "How can I change this measurement to a smaller unit?"

5. Lena found a bucket full of water in the backyard. The bucket holds 5 gallons. How many quarts of water were in the bucket?

   Ⓐ 15 quarts
   Ⓑ 20 quarts
   Ⓒ 50 quarts
   Ⓓ 1 quart

Solve another problem. As you work, ask yourself, "How can I change this measurement to a larger unit?"

6. Lena swept the front walkway. The walkway is 300 centimeters long. What is the length in meters?

   Ⓐ 200 meters
   Ⓑ 4 meters
   Ⓒ 30 meters
   Ⓓ 3 meters

Converting Customary and Metric Measures

Look at the answer choices for each question.
Read why each answer choice is correct or not correct.

5. Lena found a bucket full of water in the backyard. The bucket holds 5 gallons. How many quarts of water were in the bucket?

   Ⓐ 15 quarts

   This answer is not correct because, to change 5 gallons to quarts, you multiply 5 × 4. Since 5 × 4 = 20, 5 gallons = 20 quarts, not 15 quarts.

   ● 20 quarts

   This answer is correct because, to change 5 gallons to quarts, you multiply 5 × 4. Since 5 × 4 = 20, 5 gallons = 20 quarts.

   Ⓒ 50 quarts

   This answer is not correct because, to change 5 gallons to quarts, you multiply 5 × 4. Since 5 × 4 = 20, 5 gallons = 20 quarts, not 50 quarts.

   Ⓓ 1 quart

   This answer is not correct because, to change 5 gallons to quarts, you multiply 5 × 4. Since 5 × 4 = 20, 5 gallons = 20 quarts, not 1 quart.

6. Lena swept the front walkway. The walkway is 300 centimeters long. What is the length in meters?

   Ⓐ 200 meters

   This answer is not correct because, to change 300 centimeters to meters, you divide 300 by 100, and get 3 meters. You may have subtracted instead of divided.

   Ⓑ 4 meters

   This answer is not correct because, to change 300 centimeters to meters, you divide 300 by 100, and get 3 meters. You may have divided incorrectly.

   Ⓒ 30 meters

   This answer is not correct because, to change 300 centimeters to meters, you divide 300 by 100, and get 3 meters. You may have divided by 10 instead of 100.

   ● 3 meters

   This answer is correct because, to change 300 centimeters to meters, you divide 300 by 100, and get 3 meters.

Converting Customary and Metric Measures

PART FOUR: Learn More About Customary and Metric Measures

When you change one unit to another unit, you will not always get a whole-unit result. Sometimes, the new unit is 1 or more whole units plus a part of the unit.

▶ Think about parts of the unit when changing a smaller unit to a larger unit.

For example, 18 inches does not change easily to feet: 18 cannot be divided evenly by 12. If 1 whole foot = 12 inches, then $\frac{1}{2}$ foot = 6 inches, $\frac{1}{3}$ foot = 4 inches, and $\frac{1}{4}$ foot = 3 inches.

How many whole feet are in 18 inches? There is 1 whole foot. How many inches are left? If you subtract 12 inches from 18 inches, you have 6 inches left. This is $\frac{1}{2}$ foot. So, 18 inches = 1 foot + $\frac{1}{2}$ foot = $1\frac{1}{2}$ feet.

▶ Thinking about parts of the unit also works when changing a larger unit to a smaller unit.

To change $2\frac{1}{2}$ feet to inches, change whole and half feet separately. The 2 whole feet = 24 inches. The $\frac{1}{2}$ foot = 6 inches. So, $2\frac{1}{2}$ feet = 24 inches + 6 inches = 30 inches.

Lena's class went on a nature walk. On their walk, they made some measurements. Do Numbers 7 through 10.

7. Lena measured a fallen log. It was $20\frac{1}{2}$ feet long. What was the log's length in inches?

   Ⓐ 218 inches
   Ⓑ 206 inches
   Ⓒ 246 inches
   Ⓓ 288 inches

8. Lena measured the depth of a stream. It was $1\frac{1}{4}$ feet deep. What was its depth in inches?

   Ⓐ 15 inches
   Ⓑ 18 inches
   Ⓒ 24 inches
   Ⓓ 9 inches

9. Lena found a branch. It was 50 centimeters long. What was its length in meters?

   Ⓐ $\frac{1}{2}$ meter   Ⓒ 150 meters
   Ⓑ 1 meter   Ⓓ 25 meters

10. The students collected rocks. Later, they weighed the rocks. What was the total weight?

    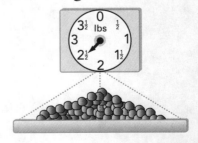

    Ⓐ 32 ounces   Ⓒ 24 ounces
    Ⓑ 16 ounces   Ⓓ 40 ounces

Lena made punch for a party. Read the recipe she used.
Then do Numbers 11 through 14.

Fizzy Fruity Punch

2 gallons orange juice $3\frac{1}{2}$ pints soda water
6 quarts cranberry juice 2 pounds ice
16 ounces apple juice 1 orange, sliced
1 cup lemon juice 1 lemon, sliced

Directions
Mix liquid ingredients together in a large bowl. Add ice and fruit slices.

11. Lena used 16 ounces of apple juice. How many quarts of apple juice did she use?
    Ⓐ 4 quarts
    Ⓑ $\frac{1}{2}$ quart
    Ⓒ $2\frac{1}{2}$ quarts
    Ⓓ 2 quarts

12. How many gallons of cranberry juice did Lena use in the punch?
    Ⓐ 6 gallons
    Ⓑ $4\frac{1}{2}$ gallons
    Ⓒ 5 gallons
    Ⓓ $1\frac{1}{2}$ gallons

13. Lena measured the soda water in cups. What was her correct measurement?
    Ⓐ 7 cups
    Ⓑ $1\frac{1}{2}$ cups
    Ⓒ $6\frac{1}{2}$ cups
    Ⓓ 3 cups

14. Lena added 4 bags of ice to the punch. How many ounces did 1 bag of ice weigh?

    Ⓐ 8 ounces
    Ⓑ 32 ounces
    Ⓒ 4 ounces
    Ⓓ 16 ounces

Converting Customary and Metric Measures

PART FIVE: Prepare for a Test

▶ A test question about customary and metric measures may ask you to use multiplication to change a measurement from a larger unit to a smaller unit.

▶ A test question about customary and metric measures may ask you to use division to change a measurement from a smaller unit to a larger unit.

Lena learned how to make spaghetti. Read what she learned. Then do Numbers 15 and 16.

How to Make Spaghetti

First, fill a large pot with 16 cups of water. Add 1 teaspoon of salt for flavor. Add 1 teaspoon of oil to keep the spaghetti from sticking together. Put the pot on the stove, and turn the burner on high heat. Heat the water until it boils. Then add the spaghetti. Use 1 pound of spaghetti for 4 people. When the water starts to boil again, set a timer for 8 minutes. After 8 minutes, remove the pot from the stove. Drain the spaghetti into a strainer in the sink. Put the spaghetti into a large bowl. Add sauce and mix well. Serve with fresh bread and a salad.

Converting Customary and Metric Measures

15. Lena and her father made spaghetti. First, she measured 16 cups of water. How many pints of water did she measure?
    - Ⓐ 4 pints
    - Ⓑ 8 pints
    - Ⓒ 64 pints
    - Ⓓ 32 pints

Converting Customary and Metric Measures

16. There are 6 people in Lena's family. She figured that she would need $1\frac{1}{2}$ pounds of spaghetti. How many ounces of spaghetti did she need?
    - Ⓐ 20 ounces
    - Ⓑ 32 ounces
    - Ⓒ 16 ounces
    - Ⓓ 24 ounces

Lena's teacher asked the students to write about their favorite place. Read what Lena wrote. Then do Numbers 17 and 18.

My Favorite Place

My favorite place is my room. It is $10\frac{1}{2}$ feet wide and 12 feet long. In my room, there is a bed and a dresser. My bed has green sheets with yellow polka dots. I have 2 windows in my room. When I look out my windows, I can see the street. Sometimes, I like to sit at my window and count the cars that go by. But the most special part of my room is the bookcase. My parents and I built it to hold all of my books. It has 5 shelves. It is 2 meters tall. I have a stool to stand on so that I can reach the books on the top shelf.

Converting Customary and Metric Measures

17. Lena's room is $10\frac{1}{2}$ feet wide. How many inches wide is her room?
    - Ⓐ 160 inches
    - Ⓑ 106 inches
    - Ⓒ 126 inches
    - Ⓓ 118 inches

Converting Customary and Metric Measures

18. Lena's bookcase is 2 meters tall. What is the height of the bookcase in centimeters?
    - Ⓐ 12 centimeters
    - Ⓑ 200 centimeters
    - Ⓒ 120 centimeters
    - Ⓓ 20 centimeters

Strategy Nine: USING ALGEBRA

PART ONE: Think About Algebra

WHAT DO YOU KNOW ABOUT ALGEBRA?

You use algebra when you look for an unknown number in a number sentence. The unknown number can be represented by a letter, such as x or n. Study the number sentence $4 + n = 6$. Add 4 to the unknown number to get the sum of 6.

The unknown number (n) is 2. $4 + 2 = 6$

▶ Write the unknown number to complete each number sentence.

a. $c + 6 = 13$ $c =$ ____
b. $4 + b = 11$ $b =$ ____
c. $8 \div 1 = m$ $m =$ ____
d. $9 - p = 4$ $p =$ ____
e. $5 \times x = 10$ $x =$ ____
f. $y \times 4 = 28$ $y =$ ____
g. $18 \div 9 = n$ $n =$ ____
h. $81 \div s = 9$ $s =$ ____

> You just reviewed how some number sentences use a letter of the alphabet to represent an unknown number.

The unknown number in a number sentence can also be represented by a box.

Study the number sentence □ + 3 = 7. Add 3 to the unknown number to get the sum of 7.

The unknown number is 4. $4 + 3 = 7$

▶ Write the unknown number to complete each number sentence.

a. 7 + □ = 15   □ = ____
b. □ ÷ 3 = 10   □ = ____
c. 17 − 8 = □   □ = ____
d. 13 − □ = 6   □ = ____
e. □ × 8 = 40   □ = ____
f. 9 × □ = 36   □ = ____
g. □ × 6 = 54   □ = ____
h. 5 × 3 = □   □ = ____

> You just reviewed how some number sentences use a box to represent an unknown number.

Using Algebra

WHAT DO YOU KNOW ABOUT PATTERNS?

You use algebra when you find the missing number in a number pattern.
Study the pattern: 2, 4, 6, 8, n
Each number in the pattern is 2 more than the number to the left: $n = 8 + 2$
So $n = 10$.

▶ What is the missing number in each pattern?
   a. 4, 7, 10, 13, ☐   ☐ = ____

   Each number in the pattern is ____ more than the number to the left.

   b. 10, 8, 6, x, 2 x = ____

   Each number in the pattern is ____ less than the number to the left.

   c. 5, ☐, 13, 17, 21   ☐ = ____

   Each number in the pattern is ____ more than the number to the left.

You use algebra when you find the missing figure in a figure pattern.
The missing figure can be represented by a line.

▶ Study the pattern △, ○, ____, △, ○, △
   Two figures form the pattern and they appear in order. The triangle is first and the circle is second. The order is repeated to complete the pattern.
   What figures goes on the line? _____

▶ Draw the missing figure in each pattern.
   a. △, ▷, _____, ◁, △

   The four figures in the series are triangles. They appear in order, and each triangle points in a different direction. The order is repeated beginning with the fifth figure in the pattern.
   The missing figure is _____.

   b. △, ▭, △, ▭, △, ▭, _____

   The next figure in the pattern is _____.

> You just reviewed information about number patterns and figure patterns.

Work with a partner.

> Individually, create two number-pattern problems, each having one missing number represented by a letter or a box. When finished, ask your partner to complete your patterns, while you complete your partner's patterns.

Using Algebra

PART TWO: Learn About Algebra

Elva made a chart to show the age her brother Roberto will be when Elva is 11. Study the chart. As you study, think about what the pattern might be.

Elva's Age	8	9	10	11
Roberto's Age	11	12	13	?

The pattern is to add 3. Roberto is 3 years older than Elva.

8 + 3 = 11   9 + 3 = 12   10 + 3 = 13   11 + 3 = 14

Roberto will be 14 when Elva is 11.

Sometimes, you are asked for the missing number or figure in a pattern. First, find the pattern. Then use the pattern to find the missing number or figure.

Number Pattern

42, 36, ___, 24, 18

What is the missing number?

42 − 6 = 36
36 − 6 = 30
30 − 6 = 24
24 − 6 = 18

Pattern: Subtract 6. The missing number is 30.

Figure Pattern

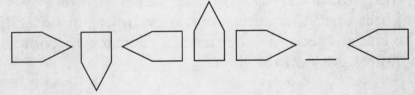

What is the missing figure?

Pattern: Each figure has turned as the hands of a clock do. The first figure points right; the next points down; the next points left; and so on. The missing figure looks like this:

You use **algebra** when you find patterns.
▶ Patterns are like rules.
▶ Use patterns to find missing numbers or figures.

Elva and her grandmother are making quilts. Study patterns A and B. Think about the pattern for each. Then do Numbers 1 through 4.

Pattern A

Elva's grandmother is making a quilt that has rows of red squares. The number of red squares in each row follows this pattern:

24, 20, 16, ___

Pattern B

Elva is making a quilt with pieces like the ones below. She is going to sew the pieces in this pattern.

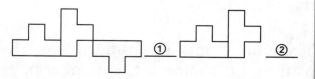

1. Look at Pattern A. What is the pattern?
   - Ⓐ Add 8.
   - Ⓑ Subtract 4.
   - Ⓒ Multiply by 4.
   - Ⓓ Subtract 8.

2. What is the missing number in Pattern A?
   - Ⓐ 21
   - Ⓑ 8
   - Ⓒ 12
   - Ⓓ 10

3. Look at Pattern B. Which of these figures belongs in space number 1?

4. Look at Pattern B. Which of these figures belongs in space number 2?

Talk about your answers to questions 1–4. Tell why you chose the answers you did.

Using Algebra

PART THREE: Check Your Understanding

Remember: You use algebra when you find patterns.

▶ Patterns are like rules.

▶ Use patterns to find missing numbers or figures.

Solve this problem. As you work, ask yourself, "What can I do to one number to find the next number in the pattern?"

Solve another problem. As you work, ask yourself, "How does the figure move from one space to the next?"

5. The number missing from the pattern is the same as the number of cousins Elva has. How many cousins does Elva have?

   35, 28, 21, ___, 7

   Ⓐ 24 cousins
   Ⓑ 17 cousins
   Ⓒ 14 cousins
   Ⓓ 28 cousins

6. Elva drew the pattern shown below. By mistake, she spilled some paint on her work. Which figure is missing?

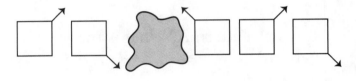

96 Using Algebra

**Look at the answer choices for each question.
Read why each answer choice is correct or not correct.**

5. The number missing from the pattern is the same as the number of cousins Elva has. How many cousins does Elva have?

    35, 28, 21, ___, 7

    Ⓐ 24 cousins

    This answer is not correct because the pattern is to subtract 7. If you subtract 7 from 21, you get 14, not 24.

    Ⓑ 17 cousins

    This answer is not correct because the pattern is to subtract 7. If you subtract 7 from 21, you get 14, not 17.

    ● 14 cousins

    This answer is correct because the pattern is to subtract 7, and 21 − 7 = 14.

    Ⓓ 28 cousins

    This answer is not correct because the pattern is to subtract 7. If you subtract 7 from 21, you get 14, not 28.

6. Elva drew the pattern shown below. By mistake, she spilled some paint on her work. Which figure is missing?

    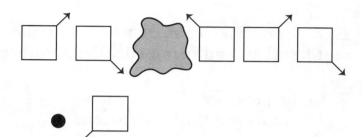

    ●

    This answer is correct because the pattern has the arrow move from corner to corner, like the hands of a clock. The third figure should point to the lower left.

    Ⓑ

    This answer is not correct because the pattern moves like the hands of a clock. The third figure should point to the lower left.

    Ⓒ

    This answer is not correct because the pattern moves like the hands of a clock. The third figure should point to the lower left.

    Ⓓ

    This answer is not correct because the pattern moves like the hands of a clock. The third figure should point to the lower left.

PART FOUR: Learn More About Algebra

You use algebra to write number sentences to solve problems.

▶ Write an addition or a subtraction sentence with the numbers given in the problem.

Use a letter of the alphabet or a box for the number you need to find. Try different numbers to find the one that works.

Elva visited the pet store with her father. She saw many different birds, reptiles, fish, and mammals. Do Numbers 7 through 10.

7. In the pet store, Elva counted 17 lizards and turtles all together. Of these animals, 12 are turtles. Which number sentence shows how many lizards there are?
   - Ⓐ $12 + 17 = n$
   - Ⓑ $12 - n = 17$
   - Ⓒ $12 + n = 17$
   - Ⓓ $29 - n = 12$

8. Two birdcages are side by side on a shelf. The length of the two cages together totals 62 inches. One cage is 24 inches long. What is the length of the bigger cage?
   - Ⓐ 36 inches
   - Ⓑ 48 inches
   - Ⓒ 24 inches
   - Ⓓ 38 inches

9. There are two cages of hamsters in the store. Elva wrote number sentences to show how many hamsters are in each cage. The □ stands for the same number of hamsters in each cage. What is the value of the □?

   Cage 1   □ + 8 = 14
   Cage 2   5 + □ = 11
   - Ⓐ 5
   - Ⓑ 6
   - Ⓒ 7
   - Ⓓ 8

10. The store sells two kinds of dog treats. There are a total of 36 boxes of dog treats, with an equal number of boxes of each kind. How many boxes of each are there?
    - Ⓐ 12 boxes
    - Ⓑ 16 boxes
    - Ⓒ 18 boxes
    - Ⓓ 24 boxes

Using Algebra

Read about Elva's camping trip. Then do Numbers 11 through 14.

Elva's family takes a camping trip every summer. When they pack, each family member has his or her own job. Elva's job is to pack the cooler. Elva packed 10 bottles of water, 15 juice boxes, 8 apples, and 6 oranges.

11. During the first two days of camping, Elva's family ate 3 apples and 2 oranges. How many pieces of fruit are left?
    Ⓐ 9 pieces
    Ⓑ 5 pieces
    Ⓒ 1 piece
    Ⓓ 11 pieces

12. The campground has 48 sites. There are 13 sites on the lake, and the rest are in the woods. Which number sentence should you use to find the number of sites in the woods?
    Ⓐ $35 - 13 = \square$
    Ⓑ $48 - 13 = \square$
    Ⓒ $48 + \square = 13$
    Ⓓ $48 + 13 = \square$

13. Elva packed equal numbers of three kinds of juice boxes—apple juice, fruit punch, and grape juice. How many of each kind of juice box did Elva pack?
    Ⓐ 3 of each kind
    Ⓑ 4 of each kind
    Ⓒ 5 of each kind
    Ⓓ 6 of each kind

14. The campground has 12 miles of hiking trails. Elva's family has hiked on 5 miles of trails. Which of these number sentences should you use to find how many miles of trails they have *not* hiked?
    Ⓐ $7 - 5 = b$
    Ⓑ $12 + 5 = b$
    Ⓒ $5 + b = 12$
    Ⓓ $12 + b = 5$

Using Algebra

PART FIVE: Prepare for a Test

▶ A test question about algebra may ask for the missing number or figure in a pattern.

▶ A test question about algebra may ask for a number sentence that solves a problem.

▶ A test question about algebra may ask for the number that completes a number sentence.

Elva is writing a report about Native Americans in the Southwest. Read the chart that she made for the report. Then do Numbers 15 and 16.

| NATIVE AMERICAN RESERVATIONS ||
State	Number of Reservations
Arizona	23
California	96
New Mexico	25
Nevada	19

Using Algebra

15. California has how many more reservations than the other three states all together?

Ⓐ 96 more reservations
Ⓑ 29 more reservations
Ⓒ 48 more reservations
Ⓓ 163 more reservations

Using Algebra

16. Elva drew this example of a pattern similar to one she saw on a Native American basket. Which of the figures would complete the pattern?

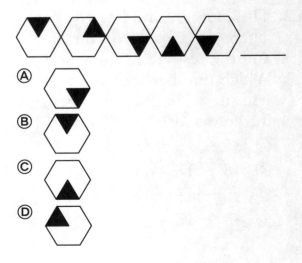

100   Using Algebra

Elva takes dance lessons and will be taking part in a recital. Read this notice that the dance teacher sent home. Then do Numbers 17 and 18.

Using Algebra

17. The opening dance at the recital will be a tap dance. The missing number in the pattern is the same as the number of dancers taking part in the opening dance. How many dancers are in the opening dance?

    4, 9, ___, 19

    Ⓐ 12 dancers
    Ⓑ 24 dancers
    Ⓒ 5 dancers
    Ⓓ 14 dancers

Using Algebra

18. The last dance of the recital will include all but 5 dancers. Which number sentence should you use to find how many dancers will take part in the last dance?

    Ⓐ 43 + ☐ = 5
    Ⓑ 43 − 5 = ☐
    Ⓒ 43 + 5 = ☐
    Ⓓ ☐ − 5 = 43

Strategies Seven–Nine REVIEW

PART ONE: Read a Story

Read the story about a special job. Then do Numbers 1 through 6.

Dan Thomas's dad is a filmmaker who has won many awards. Mr. Thomas travels all over the world, making movies about animals in the wild. Last week, he spoke about his job to the students in Dan's third-grade class.

"Filming animals in the wild is fun and exciting," he said. "I travel to wonderful places all over the world to make films. In India, I made a movie about tigers. In Australia, I made a movie about kangaroos. In China, I filmed giant pandas."

"Wild animals do not always act the way you expect them to," he said. "So, even though filming animals in the wild is fun, it can also be dangerous."

"What was the scariest thing that ever happened to you?" one of Dan's classmates asked.

Mr. Thomas smiled. "I was making a movie about African elephants. My crew and I had come upon a family of elephants taking a break in the mud. Two young calves were wrestling. The other elephants were just standing around, spraying mud on themselves. Then one of the young elephants decided to chase butterflies. It ran right toward us. Very carefully and quietly, my crew and I crawled through the mud to film the baby elephant up close."

"For more than an hour, we filmed the young elephant. It didn't seem to notice us. Suddenly, the mother elephant looked up and saw her baby and us at the same time. She came charging toward our cameras. We were lucky to get away."

After Mr. Thomas told his story about the elephants, one of the students in the class asked him how he decides on a film topic.

"Well," he explained, "I've always been very curious about nature, so I've always had lots of questions. I write those questions in my daily journal. I make films that answer those questions for me and for children like you."

Before Mr. Thomas said good-bye, he donated copies of his prize-winning films to the school video library.

▶ Converting Time and Money   ▶ Converting Customary and Metric Measures   ▶ Using Algebra

Converting Time and Money

1. If Mr. Thomas began filming the baby elephant at the time shown on the clock and continued filming for 80 minutes, what time did he stop filming?

   Ⓐ 3:46 P.M.   Ⓒ 4:04 P.M.
   Ⓑ 3:14 P.M.   Ⓓ 2:46 P.M.

Converting Customary and Metric Measures

4. The mother elephant that charged the camera crew was about 3 meters tall. What is the height of the elephant in centimeters?

   Ⓐ 13 centimeters
   Ⓑ 130 centimeters
   Ⓒ 300 centimeters
   Ⓓ 30 centimeters

Converting Time and Money

2. The usual price for one of Mr. Thomas's videos is $34.00. Which group of bills equals $34.00?

   Ⓐ two tens, one five, and six ones
   Ⓑ one ten, three fives, and ten ones
   Ⓒ two tens, two fives, and four ones
   Ⓓ three tens, one five, and one one

Using Algebra

5. The missing number in the pattern is the same as the number of elephants the crew came upon. How many elephants were there?

   5, 12, ___, 26

   Ⓐ 15 elephants   Ⓒ 7 elephants
   Ⓑ 19 elephants   Ⓓ 33 elephants

Converting Customary and Metric Measures

3. Mr. Thomas told the students that the longest elephant tusks ever measured were $11\frac{1}{2}$ feet long. What is their length in inches?

   Ⓐ 111 inches
   Ⓑ 118 inches
   Ⓒ 160 inches
   Ⓓ 138 inches

Using Algebra

6. Of the 49 films Mr. Thomas has made, 13 were about elephants. Which of these number sentences should you use to find the number of films that were *not* about elephants?

   Ⓐ 49 + ☐ = 13
   Ⓑ 13 + 49 = ☐
   Ⓒ 49 − ☐ = 13
   Ⓓ 13 − ☐ = 49

Strategies 7–9 Review

PART TWO: Read a Poster

Here is a poster that Luellen made for her report on dolphins. Read the information on the poster. Then do Numbers 7 through 12.

DOLPHINS

Dolphins are mammals that live in both warm and cool ocean waters. Some kinds of dolphins live in rivers.

A Dolphin's Body

A dolphin's body is shaped like a torpedo, larger at the front than at the back. A dolphin has

- smooth, sleek skin with very little body hair
- a back fin (Some river dolphins do not have fins.)
- a short, stiff neck
- a long nose, or beak
- a muscular tail, called a fluke, that it uses to move itself through the water
- paddle-shaped flippers

Dolphin Habits

Dolphins
- breathe air through a blowhole at the top of their head
- give birth to live young
- travel in families and in packs
- swim at high speeds
- can hold their breath underwater for 5 minutes
- are playful and like to leap, flip, and spin
- are believed to be very intelligent

Converting Time and Money

7. Luellen went to a dolphin show. The show started and stopped at the times shown on the clocks. How long was the dolphin show?

Start Stop

- Ⓐ 1 hour, 25 minutes
- Ⓑ 1 hour, 55 minutes
- Ⓒ 2 hours, 10 minutes
- Ⓓ 2 hours, 5 minutes

Converting Customary and Metric Measures

10. Because Luellen loves dolphins, she bought a small toy dolphin. She put it in a fish tank that holds 12 quarts of water. How many gallons of water does the tank hold?

| 4 quarts = 1 gallon |

- Ⓐ 6 gallons
- Ⓑ 48 gallons
- Ⓒ 24 gallons
- Ⓓ 3 gallons

Converting Time and Money

8. Luellen paid for her ticket to the dolphin show with the coins shown. What was the cost of the dolphin show?

- Ⓐ $1.05
- Ⓑ $1.60
- Ⓒ $1.65
- Ⓓ $1.80

Using Algebra

11. Which of the pictures would fill in the pattern in the bottom border of Luellen's poster?

Converting Customary and Metric Measures

9. Bottle-nosed dolphins weigh up to 440 pounds. How many ounces are in 440 pounds?

| 1 pound = 16 ounces |

- Ⓐ 55 ounces
- Ⓑ 44 ounces
- Ⓒ 7,040 ounces
- Ⓓ 2,750 ounces

Using Algebra

12. Luellen read that a sperm whale can hold its breath underwater for 75 minutes. Which of these number sentences could Luellen use to find how much longer a sperm whale can hold its breath underwater than a dolphin?

- Ⓐ $75 - 5 = x$
- Ⓒ $75 + 5 = x$
- Ⓑ $75 + x = 5$
- Ⓓ $x \times 5 = 75$

PART ONE: Think About Geometry

WHAT DO YOU KNOW ABOUT GEOMETRY?

Geometry is the study of plane figures and solid figures.
Circles, squares, and triangles are examples of plane figures.
Plane figures are flat.
Some plane figures are named for the number of sides they have.

Cubes and spheres are examples of solid figures.
Solid figures are not flat.
Solid figures have length, width, and height.
Some solid figures have edges and faces.

▶ How many sides does each plane figure have?

   a. A triangle has _____ sides.

   b. A square has _____ sides.

   c. A rectangle has _____ sides.

   d. A pentagon has _____ sides.

   e. A circle has _____ sides.

▶ How many edges does each solid figure have?

   a. A cube has _____ edges.

   b. A rectangular prism has _____ edges.

   c. A sphere has _____ edges.

   d. A square pyramid has _____ edges.

▶ How many faces does each solid figure have?

   a. A cube has _____ faces.

   b. A rectangular prism has _____ faces.

   c. A sphere has _____ faces.

   d. A square pyramid has _____ faces.

> You just reviewed information about plane figures and solid figures.

WHAT DO YOU KNOW ABOUT CONGRUENT FIGURES?

Two plane figures that have the same shape and size are congruent.

▶ Circle the two figures in each row that appear to be congruent.

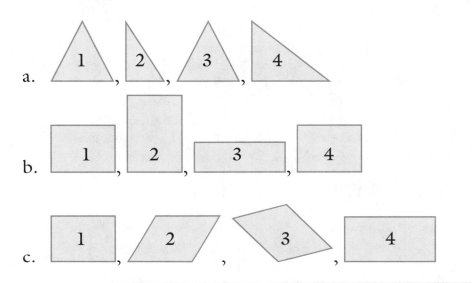

> You just reviewed information about congruent figures.

WHAT DO YOU KNOW ABOUT PERIMETER?

Perimeter is the measure of the distance around the outside of a shape or figure.

▶ On the line, write the perimeter of each figure.

a. A square measuring 4 inches on a side has a perimeter of _____ inches.

b. A rectangle with a length of 5 inches and a width of 3 inches has a perimeter of _____ inches.

c. A triangle that measures 6 centimeters on each side has a perimeter of _____ centimeters.

> You just reviewed information about perimeter.

Individually, write three perimeter problems. When finished, ask your partner to solve your problems, while you solve your partner's problems.

Using Geometry

PART TWO: Learn About Geometry

Alex's parents own a home-supply store. Study the drawing of some of the shapes that Alex sees in the store. As you study, think about how many sides, edges, or faces the figures have.

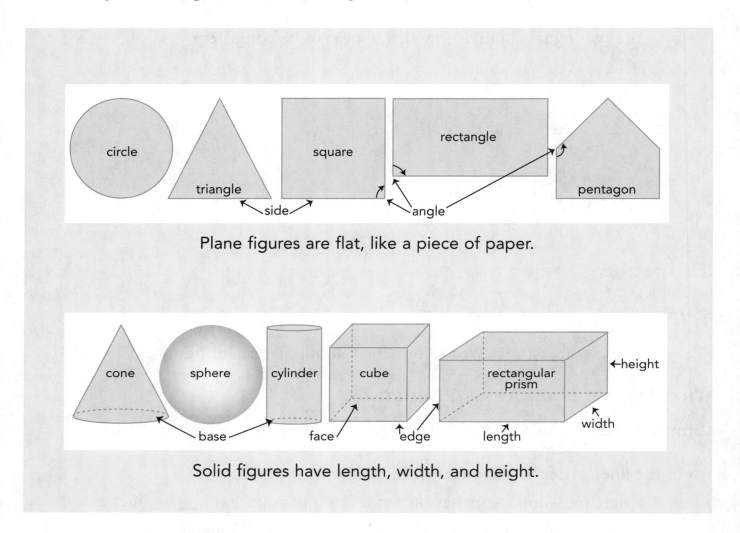

Plane figures are flat, like a piece of paper.

Solid figures have length, width, and height.

You use **geometry** when you work with figures.

▶ Plane figures are flat. Circles, squares, rectangles, and triangles are plane figures. Some plane figures are named by the number of sides they have.

▶ Solid figures are not flat. Cones, spheres, cylinders, cubes, and prisms are solid figures. Solid figures are also called space figures.

Using Geometry

Alex looked at some of the items in her parents' store. Think about the shapes of these items. Then do Numbers 1 through 4.

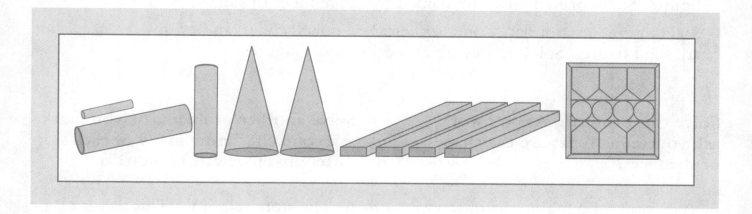

1. Look at the different figures in the stained-glass window. How many pentagons are in the window?
   - Ⓐ 6 pentagons
   - Ⓑ 8 pentagons
   - Ⓒ 4 pentagons
   - Ⓓ 5 pentagons

2. The size of a piece of lumber is stamped on 1 of the faces. What is the total number of faces on 1 piece of lumber?
   - Ⓐ 6 faces
   - Ⓑ 8 faces
   - Ⓒ 4 faces
   - Ⓓ 5 faces

3. Alex saw that some items look like solid figures. There are 3 items that look like which solid figure?
   - Ⓐ cone
   - Ⓑ rectangular prism
   - Ⓒ cylinder
   - Ⓓ cube

4. Alex counted all of the figures in the stained-glass window. There are 8 examples of which shape?
   - Ⓐ square
   - Ⓑ triangle
   - Ⓒ pentagon
   - Ⓓ circle

Talk about your answers to questions 1–4. Tell why you chose the answers you did.

Using Geometry

PART THREE: Check Your Understanding

Remember: You use geometry when you work with figures.

▶ Plane figures are flat. Circles, squares, rectangles, and triangles are plane figures. Some plane figures are named by the number of sides they have.

▶ Solid figures are not flat. Cones, spheres, cylinders, cubes, and prisms are solid figures. Solid figures are also called space figures.

Solve this problem. As you work, ask yourself, "What are the shapes of the faces?"

5. A customer bought a box of screws and a pipe. What are the names of their shapes?

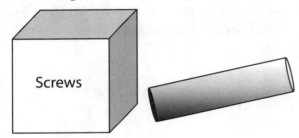

Ⓐ cube and cone
Ⓑ cube and cylinder
Ⓒ cube and sphere
Ⓓ sphere and cylinder

Solve another problem. As you work, ask yourself, "How many of each different shape can I count?"

6. The store sells TVs. This picture shows a remote control for a TV. Which shape appears the most times on the remote control?

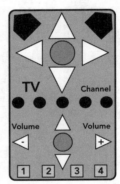

Ⓐ square
Ⓑ pentagon
Ⓒ circle
Ⓓ triangle

Look at the answer choices for each question.
Read why each answer choice is correct or not correct.

5. A customer bought a box of screws and a pipe. What are the names of their shapes?

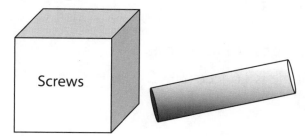

Ⓐ cube and cone

This answer is not correct because, although the box of screws is a cube, the pipe is a cylinder, not a cone.

● cube and cylinder

This answer is correct because the box of screws is a cube and the pipe is a cylinder.

Ⓒ cube and sphere

This answer is not correct because, although the box of screws is a cube, the pipe is a cylinder, not a sphere.

Ⓓ sphere and cylinder

This answer is not correct because, although the pipe is a cylinder, the box of screws is a cube, not a sphere.

6. The store sells TVs. This picture shows a remote control for a TV. Which shape appears the most times on the remote control?

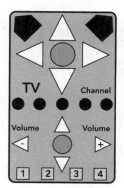

Ⓐ square

This answer is not correct because there are only 4 squares. There are more triangles and circles than squares.

Ⓑ pentagon

This answer is not correct because there are only 2 pentagons. There are more of each other kind of figure.

Ⓒ circle

This answer is not correct because, although there are 7 circles, there are 8 triangles.

● triangle

This answer is correct because there are 8 triangles, which is more than 7 circles, 4 squares, or 2 pentagons.

Using Geometry 111

PART FOUR: Learn More About Geometry

You use geometry to describe and measure figures.

▶ Figures placed in a special order make a pattern.

○ □ △ ○ □ △ This pattern can be described as circle, square, triangle.

▶ Two figures are congruent if they are the same size and shape.

These triangles are congruent.

▶ To find the perimeter of a figure, add the lengths of all sides.

The perimeter of this rectangle is 16 feet: 2 + 2 + 6 + 6 = 16.

Alex's school is having a culture day. There are displays of food, clothes, art, and other items from many different countries. Do Numbers 7 through 10.

7. Alex saw a blanket that has triangles congruent to the shaded one below. Which triangle was on the blanket?

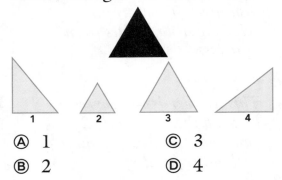

- Ⓐ 1
- Ⓑ 2
- Ⓒ 3
- Ⓓ 4

8. The pieces of a quilt on display are congruent triangles. Which of these shows pieces of the quilt?

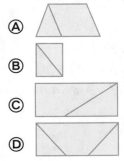

9. A display table has a pattern of shapes taped around the edges. What is the next figure in the pattern?

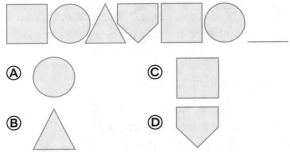

- Ⓐ circle
- Ⓑ triangle
- Ⓒ square
- Ⓓ pentagon

10. The display table is 8 feet long and 3 feet wide. What is the perimeter of the tabletop?

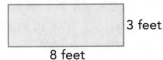

- Ⓐ 22 feet
- Ⓑ 16 feet
- Ⓒ 24 feet
- Ⓓ 11 feet

Alex visited the state capital with her family. She drew this sketch of a stone wall near the capitol building. Study the sketch. Then do Numbers 11 through 14.

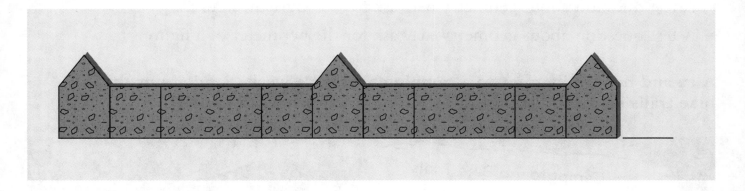

11. Alex saw that the wall has a pattern. Which of these shapes is next in the pattern?

   Ⓐ

   Ⓑ

   Ⓒ

   Ⓓ

12. The rectangles in Alex's sketch are 4 centimeters long and 2 centimeters high. What is the perimeter of 1 rectangle?
   Ⓐ 10 centimeters
   Ⓑ 12 centimeters
   Ⓒ 6 centimeters
   Ⓓ 8 centimeters

13. Alex discovered that the capitol dome was made of congruent triangles. Which of the triangles is congruent to the dark triangle?

14. The diagram below shows a garden at the statehouse. What is the perimeter of the garden?

   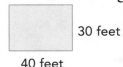

   Ⓐ 70 feet   Ⓒ 160 feet
   Ⓑ 120 feet   Ⓓ 140 feet

Using Geometry 113

PART FIVE: Prepare for a Test

- A test question about geometry may ask for the number of faces that can be seen on a solid figure.
- A test question about geometry may ask for a missing figure in a pattern.
- A test question about geometry may ask for a congruent figure.
- A test question about geometry may ask for the perimeter of a figure.

Alex and her family went to a camping area for a week. Read about the bike trails near the campgrounds. Then do Numbers 15 and 16.

Using Geometry

15. Alex took a different trail each day. The trails she took followed this pattern. What was the next trail she took?

    Ⓐ narrow trail
    Ⓑ paved path
    Ⓒ wide trail
    Ⓓ gravel road

Using Geometry

16. This dark triangle shows the shape of the campgrounds. Which of the figures is congruent to the dark triangle?

    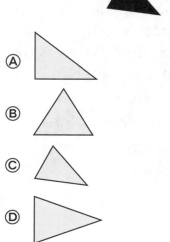

114 Using Geometry

Alex likes to visit the town center. It has statues, playgrounds, and a duck pond. Look at the picture of the statue. Then do Numbers 17 and 18.

Using Geometry

17. Alex walked all around the statue and counted the number of faces that she could see on the stand. How many faces could be seen?
    - Ⓐ 5 faces
    - Ⓑ 6 faces
    - Ⓒ 4 faces
    - Ⓓ 8 faces

Using Geometry

18. The stand of the statue is 4 feet long and 3 feet wide. What is the perimeter of the stand?
    - Ⓐ 12 feet
    - Ⓑ 7 feet
    - Ⓒ 14 feet
    - Ⓓ 10 feet

Strategy Eleven: DETERMINING PROBABILITY AND AVERAGES

PART ONE: Think About Probability and Averages

WHAT DO YOU KNOW ABOUT PROBABILITY?

Probability is the chance that an event will happen.

Probability is the number of favorable outcomes divided by the number of possible outcomes.

Some things are more likely to happen than others.

Some things are less likely to happen than others.

Some things have an equal chance of happening.

▶ Study the chart. Then read the problem and answer the questions.

Meg's Marbles	
Red	10
Blue	4
Green	2
White	4

Meg placed the four groups of marbles into a bag. Without looking, she reached into the bag and selected one marble.

a. What is the total number of marbles in Meg's bag? _____

b. What marble color is Meg most likely to select? _____

c. What marble color is Meg least likely to select? _____

d. What two marble colors does Meg have an equal chance of selecting?
_____ and _____

> You just reviewed information about probability.

What Do You Know About Averages?

An average is the sum of a group of numbers divided by the number of numbers used to find the sum.

For example, add 4, 6, 10, and 12. The sum is 32.

Divide the sum by 4, the number of numbers: 32 ÷ 4.

The quotient is the average. The average is 8.

▶ An average can be used to figure out a student's grade in a subject.
  a. Ivan's math scores are 90, 92, and 88.

     Ivan's math average is _____.

  b. Tanya's reading scores are 91, 90, and 95.

     Tanya's reading average is _____.

▶ An average can be used to figure out the number of students in each of a number of classrooms.
  a. There are 96 students in 4 third-grade classrooms.

     The average number of students per classroom is _____.

  b. Ms. Leung's 4 art classes contain 17, 20, 23, and 24 students.

     The average number of students per classroom is _____.

▶ An average can be used to figure out the daily temperature over a period of days.
  a. The temperatures at noon on 4 winter days were 30°, 32°, 35°, and 35°.

     The average noon temperature was _____.

  b. The temperatures at noon on 3 spring days were 52°, 58°, and 49°.

     The average noon temperature was _____.

> You just reviewed information about averages.

> Individually, write three problems dealing with averages. When finished, ask your partner to solve your problems while you solve your partner's problems. Together, discuss the results.

Determining Probability and Averages   117

PART TWO: Learn About Probability

Rhonda has sports cards that she keeps in a shoe box. Rhonda will close her eyes, reach into the box, and pull out a card to read. Study the types and numbers of cards. As you study, think about what card Rhonda will most likely pick.

This chart shows the number of cards that Rhonda has for each sport.

Sport	Number
Baseball	58
Football	36
Hockey	36
Basketball	5

The greatest number is 58 baseball cards. Rhonda is most likely to pick a baseball card.

The least number is 5 basketball cards. Rhonda is least likely to pick a basketball card.

You can also compare the chances of picking two different kinds of cards.

There are an equal number (36) of football and hockey cards. So, Rhonda has an equal chance of picking a football card as she has of picking a hockey card.

There are more hockey cards than basketball cards. Rhonda is more likely to pick a hockey card than a basketball card.

There are fewer football cards than baseball cards. Rhonda is less likely to pick a football card than a baseball card.

You use **probability** to find what the chance is that a certain event will happen.

▶ Some events are more likely to happen than others.

▶ Some events are less likely to happen than others.

▶ Some events have an equal chance of happening. They are equally likely.

Rhonda also has a collection of dried beans, which she keeps in a jar. Think about how many beans she has of each kind. Then do Numbers 1 through 4.

15 white beans
25 yellow beans
19 navy beans
8 black beans
6 tan beans

1. With her eyes closed, Rhonda pulls out a bean from the jar. What color of bean is she least likely to pick?
   Ⓐ white
   Ⓑ tan
   Ⓒ navy
   Ⓓ black

2. What color of bean is Rhonda less likely to pick other than a white bean?
   Ⓐ black or tan
   Ⓑ black only
   Ⓒ tan only
   Ⓓ navy

3. Rhonda made a pile of all the beans on a table. If she closes her eyes and picks up one bean, what color is she most likely to pick?
   Ⓐ white
   Ⓑ navy
   Ⓒ black
   Ⓓ yellow

4. What color of bean is Rhonda more likely to pick other than a navy bean?
   Ⓐ none
   Ⓑ black
   Ⓒ white
   Ⓓ yellow

Talk about your answers to questions 1–4. Tell why you chose the answers you did.

Determining Probability and Averages

PART THREE: Check Your Understanding

Remember: You use probability to find what the chance is that a certain event will happen.

▶ Some events are more likely to happen than others.

▶ Some events are less likely to happen than others.

▶ Some events have an equal chance of happening. They are equally likely.

Solve this problem. As you work, ask yourself, "Which item has the least number?"

5. Rhonda keeps crayons in a container. The chart shows the number of crayons of each color. If Rhonda closes her eyes and takes out a crayon, what color is she least likely to pick?

Color	Number
Blue	12
Red	10
Yellow	4
Green	9

Ⓐ blue
Ⓑ red
Ⓒ yellow
Ⓓ green

Solve another problem. As you work, ask yourself, "Which spaces are the same?"

6. Rhonda uses this board to play a game. Players roll a cube and follow the directions that the cube lands on. On which spaces is the cube equally likely to land?

Ⓐ "Lose turn" and "Win 10 points"
Ⓑ "Win 10 points" and "Pick card"
Ⓒ "Win 10 points" and "Lose 10 points"
Ⓓ "Lose turn" and "Lose 10 points"

Look at the answer choices for each question.
Read why each answer choice is correct or not correct.

5. Rhonda keeps crayons in a container. The chart shows the number of crayons of each color. If Rhonda closes her eyes and takes out a crayon, what color is she least likely to pick?

Color	Number
Blue	12
Red	10
Yellow	4
Green	9

Ⓐ blue

This answer is not correct because there are more blue crayons than any other color of crayon. So, Rhonda is more likely to pick blue.

Ⓑ red

This answer is not correct because there are fewer yellow crayons than red crayons. So, Rhonda is not less likely to pick a red crayon than a yellow one.

● yellow

This answer is correct because there are fewer yellow crayons than any other color of crayon. So, Rhonda is least likely to pick yellow.

Ⓓ green

This answer is not correct because there are fewer yellow crayons than green crayons. So, Rhonda is not less likely to pick a green crayon than a yellow one.

6. Rhonda uses this board to play a game. Players roll a cube and follow the directions that the cube lands on. On which spaces is the cube equally likely to land?

Ⓐ "Lose turn" and "Win 10 points"

This answer is not correct because the "Lose turn" space is larger than the "Win 10 points" space. So, a cube is more likely to land on "Lose turn."

Ⓑ "Win 10 points" and "Pick card"

This answer is not correct because the "Pick card" space is larger than the "Win 10 points" space. So, a cube is more likely to land on "Pick card."

● "Win 10 points" and "Lose 10 points"

This answer is correct because the "Win 10 points" space and the "Lose 10 points" space are the same size. So, a cube is equally likely to land on either of these spaces.

Ⓓ "Lose turn" and "Lose 10 points"

This answer is not correct because the "Lose turn" space is larger than the "Lose 10 points" space. So, a cube is more likely to land on "Lose turn."

Determining Probability and Averages

PART FOUR: Learn More About Averages

▶ You can find the **average** of a group of numbers.

Find the total number of items in all the groups.

Put the total number of items into equal groups. Make the same number of groups that there were at the start. (Divide the total by the number of groups.)

Rhonda played baseball last spring. The chart shows how many hits she got each month. Study the chart. Then do Numbers 7 through 10.

Month	Number of Hits
April	8
May	12
June	10

7. What is the total number of hits that Rhonda got in April, May, and June?
   - Ⓐ 12 hits
   - Ⓑ 22 hits
   - Ⓒ 30 hits
   - Ⓓ 10 hits

8. How many groups (or months) are shown in the chart?
   - Ⓐ 3 groups
   - Ⓑ 4 groups
   - Ⓒ 8 groups
   - Ⓓ 12 groups

9. What is the average number of hits that Rhonda got during baseball season?
   - Ⓐ 15 hits
   - Ⓑ 30 hits
   - Ⓒ 8 hits
   - Ⓓ 10 hits

10. Which of these pictures show the total number of hits in equal groups?

    Ⓐ

    Ⓑ

    Ⓒ

    Ⓓ

Determining Probability and Averages

Rhonda's family goes camping at Brewster Forest. Look at the maps of three hiking trails. Then do Numbers 11 through 14.

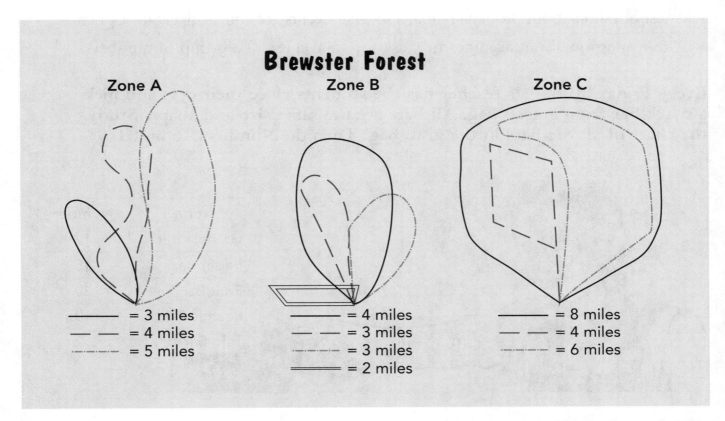

11. What is the total length of the trails in Zone A?
    - Ⓐ 10 miles
    - Ⓑ 8 miles
    - Ⓒ 12 miles
    - Ⓓ 7 miles

12. There are 3 trails in Zone A. What is the average length of a trail in Zone A?
    - Ⓐ 12 miles
    - Ⓑ 3 miles
    - Ⓒ 4 miles
    - Ⓓ 2 miles

13. Over 4 days, Rhonda's family hiked on one of each trail in Zone B. What is the average daily distance they hiked?
    - Ⓐ 3 miles
    - Ⓑ 4 miles
    - Ⓒ 5 miles
    - Ⓓ 2 miles

14. What is the average length of a trail in Zone C?
    - Ⓐ 18 miles
    - Ⓑ 4 miles
    - Ⓒ 5 miles
    - Ⓓ 6 miles

Determining Probability and Averages

PART FIVE: Prepare for a Test

▶ A test question about probability may ask what event is most likely to happen.
▶ A test question about probability may ask what event is least likely to happen.
▶ A test question about probability may ask what events are equally likely to happen.
▶ A test question about averages may ask for an average of a group of numbers.

Every Friday, Rhonda's teacher has the students close their eyes and pick a snack bar from a grab bag. All bars are the same size and shape. Study the chart of all the bars in the grab bag. Then do Numbers 15 and 16.

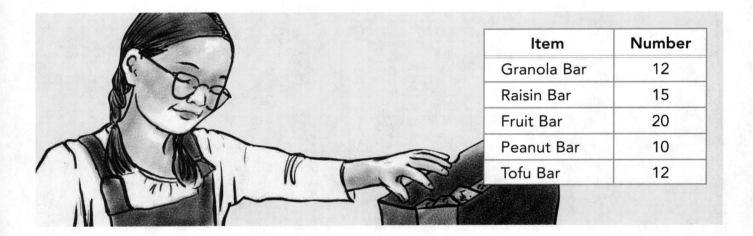

Item	Number
Granola Bar	12
Raisin Bar	15
Fruit Bar	20
Peanut Bar	10
Tofu Bar	12

Determining Probability and Averages

15. Rhonda is the first student to pick a snack bar. What bar is she most likely to pick?
    Ⓐ raisin bar
    Ⓑ peanut bar
    Ⓒ fruit bar
    Ⓓ tofu bar

Determining Probability and Averages

16. Rhonda is the first student to pick a snack bar. Which of these statements is true?
    Ⓐ Rhonda is least likely to pick a tofu bar.
    Ⓑ Rhonda is equally likely to pick a raisin bar or a tofu bar.
    Ⓒ Rhonda is least likely to pick a fruit bar.
    Ⓓ Rhonda is equally likely to pick a granola bar or a tofu bar.

124 Determining Probability and Averages

Rhonda's class is collecting bottles and cans. Study the chart of the number of bottles and cans that they collected each day. Then do Numbers 17 and 18.

Day	Bottles	Cans
Monday	5	6
Tuesday	3	3
Wednesday	4	4
Thursday	3	7
Friday	5	5

Determining Probability and Averages

17. What is the average number of bottles that Rhonda's class collected daily?
    - Ⓐ 5 bottles
    - Ⓑ 6 bottles
    - Ⓒ 4 bottles
    - Ⓓ 20 bottles

Determining Probability and Averages

18. What is the average number of cans that Rhonda's class collected daily?
    - Ⓐ 10 cans
    - Ⓑ 4 cans
    - Ⓒ 5 cans
    - Ⓓ 25 cans

Strategy Twelve: INTERPRETING GRAPHS AND CHARTS

PART ONE: Think About Graphs and Charts

WHAT DO YOU KNOW ABOUT GRAPHS AND CHARTS?

A graph can be a bar graph or a line graph.

A bar graph uses bars and numbers to show how many.

A line graph uses lines and points to show how many.

A chart uses numbers to show how many.

▶ Study the bar graph and the line graph. They both show the same information about apples picked by Tyla. Then answer the questions.

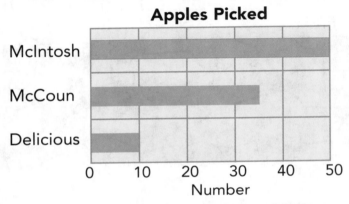
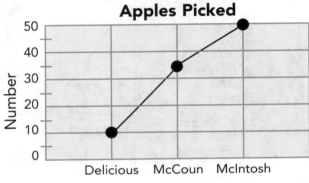

a. How many Delicious apples did Tyla pick? _____

b. How many McCoun apples did Tyla pick? _____

c. How many McIntosh apples did Tyla pick? _____

▶ Study the chart. It shows the same information that appears in the graphs above.

Apples Picked	
Kind	Number
McIntosh	50
McCoun	35
Delicious	10

a. How many more McIntosh apples did Tyla pick than McCoun? _____

b. How many more McCoun apples did Tyla pick than Delicious? _____

c. What is the total number of apples Tyla picked? _____

> You just reviewed information about bar graphs, line graphs, and charts.

126 Interpreting Graphs and Charts

WHAT ADDITIONAL INFORMATION DO YOU KNOW ABOUT GRAPHS AND CHARTS?

Graphs and charts can be used to organize information in rows and columns. Graphs that organize information in rows and columns are called grids. A chart that organizes days and dates in rows and columns is called a calendar.

▶ Study the grid that shows the layout of the Wilson property. Then answer the questions. The first question has been completed for you.

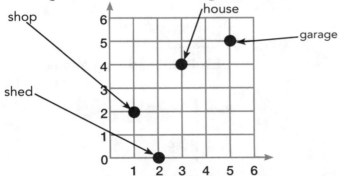

a. Where is the garage located on the grid? <u>over 5</u> and <u>up 5</u>.
b. Where is the shed located on the grid? _____ and _____.
c. Where is the house located on the grid? _____ and _____.
d. What is located at the point that is over 1 and up 2? _____
e. What is located at the point that is over 5 and up 5? _____

▶ Study the calendar. Then answer the questions. The first question has been completed for you.

			JUNE			
Sun.	Mon.	Tue.	Wed.	Thur.	Fri.	Sat.
		1	2	3	4	5
6	7	8	9	10	11	12
13	14	15	16	17	18	19
20	21	22	23	24	25	26
27	28	29	30			

a. What date is in the third row, fourth column? <u> 16 </u>
b. What date is in the fourth row, seventh column? _____
c. What day and date is the last day of the month? _____, _____
d. In what row and column does June 21 appear? row _____, column _____
e. In what row and column does June 9 appear? row _____, column _____

> You just reviewed information about a grid and a calendar.

 **Work with a partner.** Individually, create two grids and write two questions about each grid. When finished, ask your partner to solve your problems while you solve your partner's problems. Together, discuss the results.

Interpreting Graphs and Charts

PART TWO: Learn About Graphs and Charts

Tani visited a fish hatchery, where fish are raised. Later, the fish are put into lakes in the state. He made some graphs and charts to show the numbers of fish put into the lakes last week. Study the graphs and charts. As you study, think about how the information is shown.

Fish Put into Lakes Last Week	
Kind	Number
Bass	80
Pike	40
Trout	60
Salmon	50

You can show this same information in a bar graph.

Look for the number that lines up with the top of each bar. The salmon bar is halfway between 40 and 60, so it stands for 50. The other bars line up exactly with the numbers on the bottom of the graph.

A line graph can also show amounts. Line graphs show changes in amounts over a period of time.

Look at the point above each week on the graph. Then look at the numbers on the left side of the graph. The numbers beside each point show what the points on the graph stand for.

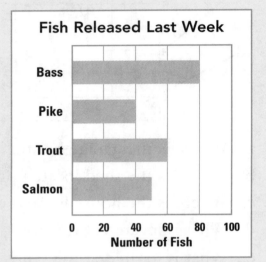

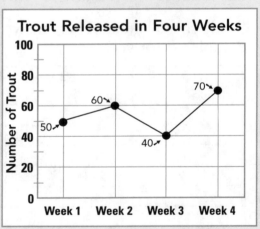

You use **graphs and charts** to show information.
- A chart uses numbers to show how many.
- A bar graph uses bars and numbers to show how many.
- A line graph uses lines, points, and numbers to show how many.

128 Interpreting Graphs and Charts

Tani found out more about fish. Think about how the chart and the graph show the information. Then do Numbers 1 through 4.

Group	Number of Species
Salmon and Trout	1,000
Codfish	800
Flatfish	500
Catfish	2,500

The chart above shows about how many species there are for four fish groups.

The graph to the right shows the maximum, or greatest, length of four kinds of fish.

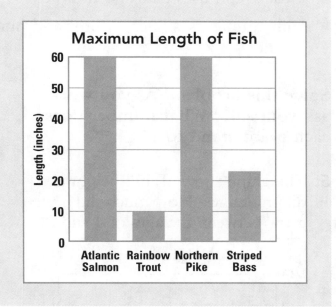

1. Which of these statements about the chart is true?
   Ⓐ The group with the most species is salmon and trout.
   Ⓑ The group with the fewest species is codfish.
   Ⓒ There are about 300 more species of codfish than flatfish.
   Ⓓ There are about 1,000 fewer species of salmon and trout than catfish.

2. Which of these fish groups has the most species?
   Ⓐ salmon and trout
   Ⓑ flatfish
   Ⓒ codfish
   Ⓓ catfish

3. Most northern pike are shorter in length than what measurement?
   Ⓐ 10 inches
   Ⓑ 50 inches
   Ⓒ 60 feet
   Ⓓ 60 inches

4. Which of these statements about the bar graph is true?
   Ⓐ Atlantic salmon and northern pike are about the same length.
   Ⓑ The Atlantic salmon is the longest of all the fish shown on the graph.
   Ⓒ Most rainbow trout are longer than 20 inches.
   Ⓓ Most striped bass are shorter than 10 inches.

Talk about your answers to questions 1–4. Tell why you chose the answers you did.

Interpreting Graphs and Charts

PART THREE: Check Your Understanding

Remember: You use graphs and charts to show information.

▶ A chart uses numbers to show how many.

▶ A bar graph uses bars and numbers to show how many.

▶ A line graph uses lines, points, and numbers to show how many.

Solve this problem. As you work, ask yourself, "What number does each point stand for?"

5. The graph shows Tani's height at different ages. How much did Tani grow between the ages of 4 and 8?

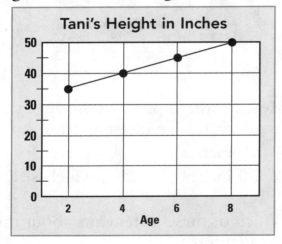

Ⓐ 5 inches
Ⓑ 10 inches
Ⓒ 20 inches
Ⓓ 15 inches

Solve another problem. As you work, ask yourself, "What number is shown for each age?"

6. The chart shows Tani's weight at different ages. How much weight did Tani gain from age 2 to age 8?

| Tani's Weight ||
Age	Weight
2	29 lb
4	38 lb
6	51 lb
8	60 lb

Ⓐ 21 lb
Ⓑ 9 lb
Ⓒ 31 lb
Ⓓ 60 lb

130 Interpreting Graphs and Charts

Look at the answer choices for each question.
Read why each answer choice is correct or not correct.

5. The graph shows Tani's height at different ages. How much did Tani grow between the ages of 4 and 8?

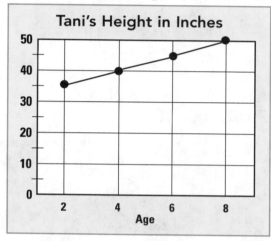

Ⓐ 5 inches

This answer is not correct because at age 8, Tani was 50 inches; and at age 4, he was 40 inches: 50 − 40 does not equal 5.

● 10 inches

This answer is correct because Tani's height at age 8 minus his height at age 4 is 50 in. − 40 in. = 10 in.

Ⓒ 20 inches

This answer is not correct because at age 8, Tani was 50 inches. If you subtract 20 inches from 50 inches, you get 30 inches, which is less than his height even at age 2.

Ⓓ 15 inches

This answer is not correct because 15 inches is how much Tani grew from age 2 through age 8: 50 in. − 35 in. = 15 in.

6. The chart shows Tani's weight at different ages. How much weight did Tani gain from age 2 to age 8?

Tani's Weight	
Age	Weight
2	29 lb
4	38 lb
6	51 lb
8	60 lb

Ⓐ 21 lb

This answer is not correct because 60 lb − 29 lb = 31 lb, not 21 lb.

Ⓑ 9 lb

This answer is not correct because 9 lb is the difference in Tani's weight only from age 6 to age 8: 60 lb − 51 lb = 9 lb.

● 31 lb

This answer is correct because 31 lb is the difference in Tani's weight from age 2 to age 8: 60 lb − 29 lb = 31 lb.

Ⓓ 60 lb

This answer is not correct because 60 lb was Tani's weight at age 8.

Interpreting Graphs and Charts

PART FOUR: Learn More About Graphs and Charts

You use **graphs and charts** to organize information in rows and columns.

▶ Some graphs, or grids, have points that can be described by location.

In this grid, point *D* is over 4 and up 2. Point *B* is over 2 and up 3.

▶ Calendars show days and numbers in rows and columns.

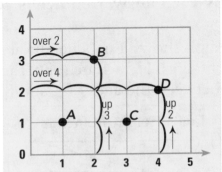

Tani's uncle owns a nursery. Many of the trees and shrubs are planted in rows. Look at the grid. Then do Numbers 7 through 10.

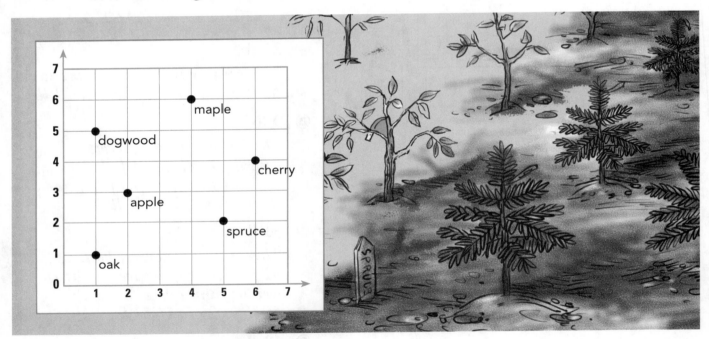

7. What tree is over 2 and up 3?
   - Ⓐ oak tree
   - Ⓑ apple tree
   - Ⓒ spruce tree
   - Ⓓ cherry tree

8. Where is the dogwood tree?
   - Ⓐ over 1 and up 5
   - Ⓑ over 5 and up 1
   - Ⓒ over 2 and up 5
   - Ⓓ over 5 and up 2

9. Where is the spruce tree?
   - Ⓐ over 3 and up 5
   - Ⓑ over 2 and up 5
   - Ⓒ over 5 and up 3
   - Ⓓ over 5 and up 2

10. What tree is over 4 and up 6?
    - Ⓐ apple tree
    - Ⓑ cherry tree
    - Ⓒ dogwood tree
    - Ⓓ maple tree

132 Interpreting Graphs and Charts

Read Tani's baseball schedule. Then do Numbers 11 through 14.

Baseball Schedule

The opening-day parade will be held on the first Sunday in May.

Practice begins every Wednesday at 5:00 P.M.

Schedule for first 6 games:
- May 9 at 5:30
- May 11 at 1:00
- May 16 at 5:30
- May 17 at 3:15
- May 24 at 5:15
- May 31 at 5:30

MAY

SUN	MON	TUES	WED	THURS	FRI	SAT
				1	2	3
4	5	6	7	8	9	10
11	12	13	14	15	16	17
18	19	20	21	22	23	24
25	26	27	28	29	30	31

11. What is the date of the opening-day parade?
    - Ⓐ May 1
    - Ⓑ May 3
    - Ⓒ May 4
    - Ⓓ May 11

12. On which of these dates does Tani have baseball practice?
    - Ⓐ May 6
    - Ⓑ May 15
    - Ⓒ May 20
    - Ⓓ May 28

13. On which of these days will Tani have his first baseball game?
    - Ⓐ Friday
    - Ⓑ Saturday
    - Ⓒ Sunday
    - Ⓓ Wednesday

14. On which of these days is the last game of the month?
    - Ⓐ Friday
    - Ⓑ Saturday
    - Ⓒ Sunday
    - Ⓓ Tuesday

Interpreting Graphs and Charts

PART FIVE: Prepare for a Test

▶ A test question about graphs and charts may ask for information from a bar graph or a chart.

▶ A test question about graphs may ask for information from a line graph.

▶ A test question about graphs may ask for information from a grid.

▶ A test question about charts may ask for information from a calendar.

Every spring, Tani helps his mother plant a garden. Study the graphs. Then do Numbers 15 and 16.

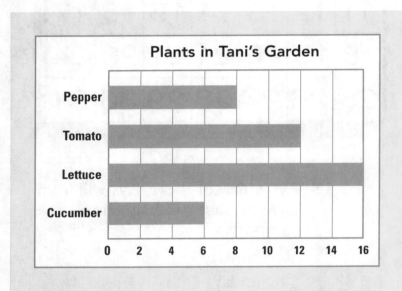
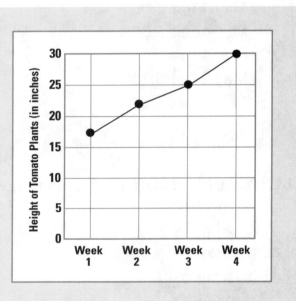

Interpreting Graphs and Charts

15. How many more lettuce plants than tomato plants are in the garden?
   Ⓐ 8 lettuce plants
   Ⓑ 14 lettuce plants
   Ⓒ 16 lettuce plants
   Ⓓ 4 lettuce plants

Interpreting Graphs and Charts

16. Tani kept track of the growth of one tomato plant. How much did the plant grow from week 3 to week 4?
   Ⓐ 42 inches
   Ⓑ 30 inches
   Ⓒ 5 inches
   Ⓓ 3 inches

Tani also plants his own garden. This grid shows where Tani put his plants. Study the grid. Then do Numbers 17 and 18.

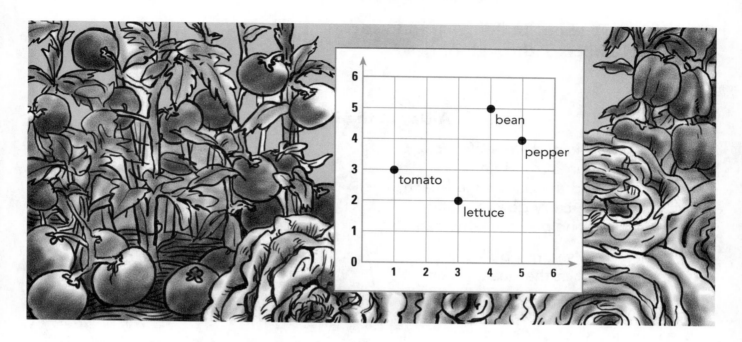

Interpreting Graphs and Charts

17. Which of these plants is over 4 and up 5 on the grid?
    - Ⓐ pepper
    - Ⓑ tomato
    - Ⓒ bean
    - Ⓓ lettuce

Interpreting Graphs and Charts

18. On the third Saturday in May, Tani planted a row of beans. He wants to plant another row of beans 3 weeks after this date. On what date should Tani plant the second row of beans?

    - Ⓐ May 31
    - Ⓑ June 7
    - Ⓒ May 1
    - Ⓓ June 8

Interpreting Graphs and Charts 135

Strategies Ten–Twelve REVIEW

PART ONE: Read a Story

Read the story about building a sandcastle. Then do Numbers 1 through 6.

A Day at the Beach

Larry was excited when he woke up. It was Saturday, and he was going to spend the day at the beach with his family. Larry got up, had breakfast, and helped pack the car.

When Larry's family pulled into the parking lot at the beach, they saw a sign: SANDCASTLE CONTEST TODAY! Larry said excitedly, "I'm glad we brought all of our sand toys."

Larry and his sisters went right to work. They collected shells and other objects to use as decorations. Larry collected 6 beautiful shells. His sister Brianne found 5 shells in many different colors. Larry's other sister, Dawn, picked up 4 more shells for their collection. They spread out the decorations in the sand so that they could study what they had found.

The whole family helped build the castle. They even made a moat and filled it with water. The final touches included shells, rocks, and seaweed decorations. Larry was sure they would win the contest.

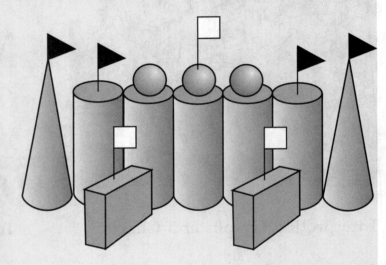

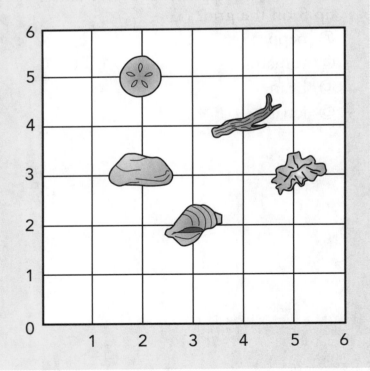

▶ Using Geometry ▶ Determining Probability and Averages ▶ Interpreting Graphs and Charts

Using Geometry

1. Which of these shapes appears only four times in the picture?
   - Ⓐ rectangle
   - Ⓑ square
   - Ⓒ cone
   - Ⓓ triangle

Using Geometry

2. Which of these solid figures appears only two times in the picture?
   - Ⓐ cube
   - Ⓑ cone
   - Ⓒ sphere
   - Ⓓ cylinder

Determining Probability and Averages

3. What is the average number of shells that Larry and his 2 sisters each collected for the sandcastle?
   - Ⓐ 15 shells
   - Ⓑ 3 shells
   - Ⓒ 5 shells
   - Ⓓ 6 shells

Determining Probability and Averages

4. Larry's family brought a bag of apples to the beach. There were 2 yellow apples, 4 red apples, and 6 green apples. If Larry reaches into the bag without looking, what color of apple is he least likely to pick?
   - Ⓐ a red apple
   - Ⓑ a yellow apple
   - Ⓒ a green apple
   - Ⓓ a yellow or a green apple

Interpreting Graphs and Charts

5. What item on the grid is over 3 and up 2?
   - Ⓐ shell
   - Ⓑ seaweed
   - Ⓒ stick
   - Ⓓ sand dollar

Interpreting Graphs and Charts

6. This chart shows the number of different kinds of shells that Larry and his family collected all day. Together how many shells did they find?

Kind of Shell		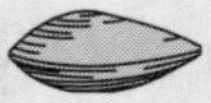	
Number	22	7	18

   - Ⓐ 22 shells
   - Ⓑ 37 shells
   - Ⓒ 40 shells
   - Ⓓ 47 shells

Strategies 10–12 Review

PART TWO: Read Instructions

Read the instructions for creating a craft project. Then do Numbers 7 through 12.

Handicraft Kit

You can use this kit to make and decorate many fun projects.

- Use the stencils to color or paint designs and patterns.
- Use pom-poms to brighten up your life.
- Use your imagination to design your own works of art!

This graph shows all of the pieces included in the Handicraft Kit.

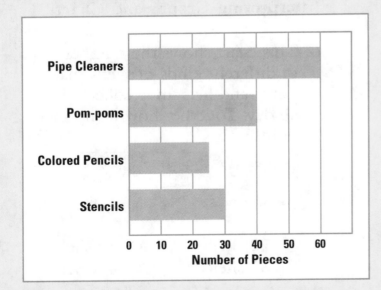

This line graph shows about how long each of four different projects will take.

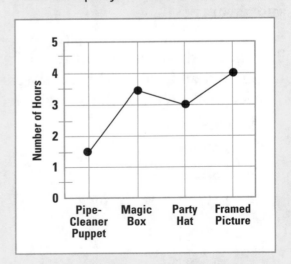

Using Geometry

7. The pattern shows several triangles like this one.

Which of these triangles is congruent to the dark triangle?

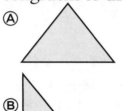
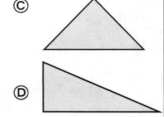

Determining Probability and Averages

10. The instructions show that the kit includes colored pencils. There is 1 black, 1 brown, 1 white, and 1 yellow pencil. There are also 6 blue, 4 red, 3 purple, 5 green, and 3 pink. If you were to close your eyes and reach into the box, which of these colors would you most likely pick?

- Ⓐ blue
- Ⓑ red
- Ⓒ green
- Ⓓ pink

Using Geometry

8. Look at the pattern shown in the instructions. If one more square were added to the right side of the pattern, what square would it be?

Interpreting Graphs and Charts

11. How many hours in all would it take to make a magic box and a party hat?

- Ⓐ 6 hours
- Ⓑ 6½ hours
- Ⓒ ½ hour
- Ⓓ 5 hours

Determining Probability and Averages

9. Look at the line graph included in the instructions. What is the average time needed to complete 1 of the 4 projects?

- Ⓐ 4 hours
- Ⓑ 3 hours
- Ⓒ 12 hours
- Ⓓ 3½ hours

Interpreting Graphs and Charts

12. How many pieces are there all together in the Handicraft Kit?

- Ⓐ 145 pieces
- Ⓑ 160 pieces
- Ⓒ 150 pieces
- Ⓓ 155 pieces

Strategies One–Twelve FINAL REVIEW

PART ONE: Read a Folktale

This Russian folktale tells about the importance of teamwork. Read the folktale. Then do Numbers 1 through 12.

The Amazing Turnip

A man once planted a turnip seed in his vegetable garden.

The man, who had a wife, a son, a daughter-in-law, 5 grandchildren, a dog, and a cat, tended his garden every day. He took extra-special care of his turnip plant. He gave it extra water and extra room to grow. And grow it did. It grew and grew and grew until it was so big its leaves shaded the whole garden.

It was time to pull the turnip out of the ground. So the man grabbed the turnip by its leaves and pulled. He pulled harder and harder. But no matter how hard he pulled, the turnip would not budge.

The man's wife came to help. She wrapped her arms around her husband, and together they pulled. They pulled harder and harder. But no matter how hard they pulled, the turnip would not budge.

The man's son and daughter-in-law came to help. The son wrapped his arms around his mother. The daughter-in-law wrapped her arms around her husband. Together they pulled. They pulled harder and harder. But no matter how hard they pulled, the turnip would not budge.

The man's 5 grandchildren came to help. They grabbed on to their mother. Together they pulled. They pulled harder and harder. But no matter how hard they pulled, the turnip would not budge.

Then the man's dog came to help. The turnip would not budge.

Next came the cat. The turnip would not budge.

Finally, a tiny mouse came to help.

The tiny mouse wrapped itself around the cat. Together, the mouse, the cat, the dog, the man's 5 grandchildren, his daughter-in-law, his son, his wife, and the man himself pulled. And the gigantic turnip came flying out of the ground. The man and all his helpers fell in a big heap. The turnip was so large, the family had enough to eat for a year.

Building Number Sense

1. The weight of the turnip was 5,248 ounces. What is the value of the 2 in 5,248?
   - Ⓐ 2
   - Ⓑ 20
   - Ⓒ 200
   - Ⓓ 2,000

Using Estimation

2. The village grocer offered the man $4.57 for 100 pounds of turnip. To the nearest 10¢, how much money would the grocer have to pay for 300 pounds of turnip?
   - Ⓐ $13.00
   - Ⓑ $13.80
   - Ⓒ $14.00
   - Ⓓ $13.30

Applying Addition

3. Two weeks before it was pulled from the ground, the turnip was 283 centimeters tall. In those two weeks, it grew 81 centimeters more. How tall was the turnip when it was pulled from the ground?
   - Ⓐ 364 centimeters
   - Ⓑ 264 centimeters
   - Ⓒ 202 centimeters
   - Ⓓ 352 centimeters

Applying Subtraction

4. The man's garden was 420 centimeters wide. The width of the turnip was 139 centimeters. What was the difference between the widths of the garden and the turnip?
   - Ⓐ 381 centimeters
   - Ⓑ 290 centimeters
   - Ⓒ 359 centimeters
   - Ⓓ 281 centimeters

Applying Multiplication

5. During the last week of its growth, the turnip grew 13 pounds a day. How many pounds did the turnip grow that week?
   - Ⓐ 91 pounds
   - Ⓑ 83 pounds
   - Ⓒ 96 pounds
   - Ⓓ 80 pounds

Applying Division

6. The man cut the turnip into 75 large chunks. He shared the turnip chunks equally among his family and 4 other families in the neighborhood. How many turnip chunks did each family get?
   - Ⓐ 25 turnip chunks
   - Ⓑ 21 turnip chunks
   - Ⓒ 15 turnip chunks
   - Ⓓ 18 turnip chunks

Strategies 1–12 Final Review

Converting Time and Money

7. The man began to pull the turnip at the time shown on the clock. The turnip came flying out of the ground 82 minutes later. What time was the turnip pulled from the ground?

- Ⓐ 10:55 A.M.
- Ⓒ 11:03 A.M.
- Ⓑ 12:45 P.M.
- Ⓓ 11:45 A.M.

Using Geometry

10. The man built a cart to wheel the turnip to his house. The floor of the cart was 6 feet long and 7 feet wide. What was the perimeter of the floor of the cart?

- Ⓐ 13 feet
- Ⓑ 42 feet
- Ⓒ 19 feet
- Ⓓ 26 feet

Converting Customary and Metric Measures

8. The man's wife made $2\frac{1}{2}$ gallons of turnip stew for Sunday dinner. How many quarts of stew did she make?

> 1 gallon = 4 quarts

- Ⓐ 8 quarts
- Ⓑ 10 quarts
- Ⓒ 16 quarts
- Ⓓ 32 quarts

Determining Probability and Averages

11. What is the average age of the man's grandchildren?

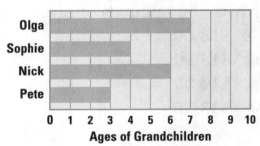

- Ⓐ 4 years
- Ⓒ 3 years
- Ⓑ 5 years
- Ⓓ 6 years

Using Algebra

9. The missing number in the pattern is the same as the number of turnip meals the man and his wife made in August. How many turnip meals did they make in August?

7, 15, ___, 31

- Ⓐ 23 meals
- Ⓒ 8 meals
- Ⓑ 39 meals
- Ⓓ 21 meals

Interpreting Graphs and Charts

12. Look at the graph in Problem 11. What is the difference in years between the oldest and the youngest grandchild?

- Ⓐ 4 years
- Ⓑ 8 years
- Ⓒ 5 years
- Ⓓ 3 years

PART TWO: Read a Story

**Read the story about Lian and her friend Kameko.
Then do Numbers 13 through 24.**

Best Friends

When Lian broke her leg in an accident, she had to spend several weeks at home. Each day after school, her best friend, Kameko, came to visit. They talked and played games, like Scribble, Polymono, and Yahoo. While they played, Lian tried not to complain or to worry.

After a few days, though, Kameko's friendly smile could not cheer Lian. "What if my leg does not heal properly?" she said. "What if I can never dance ballet again?"

"It will heal. You will dance," Kameko comforted her friend.

Lian shook her head sadly. "Maybe," she whispered, "and maybe not."

Later that day, Kameko overheard a conversation between her mother and Lian's mother. "The doctor told us that Lian's leg will heal," Lian's mother had said. "But Lian is so unhappy. I believe her unhappiness is slowing her recovery."

It had been Kameko's goal to help Lian get better quickly, but Kameko had run out of ways to boost Lian's spirits.

"Why don't you and Lian play droodles?" Kameko's father said. "A droodle is a combination of a riddle, a scribble, and a doodle. Roger Price published the first book of droodles in the 1950s. Here, I'll show you. I'll draw a droodle. Then you guess what it is."

"It looks like a squished hat," said Kameko.

"Good guess," said her father. "Actually, it's a turtle on a skateboard."

The next day, and every day until Lian's leg was completely healed, Kameko and Lian created droodles. "Why don't we make our own book of droodles?" Lian said. So they did. Lian, Kameko, Lian's mom, and both of Kameko's parents all drew droodles. By the time they put their book of droodles together, Lian's leg was healed. Not long after that, she was dancing in her ballet recital.

Name	Number
Lian	12
Kameko	14
Lian's Mom	22
Kameko's Mom	14
Kameko's Dad	10

This chart shows how many droodles each person created for Lian's and Kameko's book of droodles.

Building Number Sense

13. Lian's total score for 5 games of Scribble was more than 1,000 points. The exact score was an even number. Which of these could be Lian's exact score for 5 games?

- Ⓐ 1,154
- Ⓑ 1,261
- Ⓒ 1,083
- Ⓓ 1,149

Using Estimation

14. Lian and Kameko plan to sell copies of their droodles book for $1.42. To the nearest 10¢, what would a copy of the book cost?

- Ⓐ $1.00
- Ⓑ $1.30
- Ⓒ $1.40
- Ⓓ $1.50

Applying Addition

15. The list shows how many droodles were on the first 3 pages of the girls' book. How many droodles were on the 3 pages?

   Page 1   5 droodles
   Page 2   4 droodles
   Page 3   6 droodles

- Ⓐ 17 droodles
- Ⓑ 25 droodles
- Ⓒ 21 droodles
- Ⓓ 15 droodles

Applying Subtraction

16. The girls played 126 games of Yahoo during Lian's recovery. Lian won 74 games. How many games did Kameko win?

- Ⓐ 50 games
- Ⓑ 52 games
- Ⓒ 65 games
- Ⓓ 62 games

Applying Multiplication

17. In one Scribble game, Lian got 3 times the *total* letter value of the word *fast*. The total letter value was 22 points. What was Lian's correct score for *fast*?

- Ⓐ 55 points
- Ⓑ 66 points
- Ⓒ 35 points
- Ⓓ 46 points

Applying Division

18. Each of the last 7 pages of the girls' droodles book had the same number of droodles. If there were 42 droodles on these pages, how many were on each page?

- Ⓐ 8 droodles
- Ⓑ 5 droodles
- Ⓒ 6 droodles
- Ⓓ 7 droodles

Strategies 1–12 Final Review

Converting Time and Money

19. Lian's mom bought Lian Polymono as a get-well gift. Lian's mom paid exactly $39.00 for the game. What bills did she use?

- Ⓐ three tens, two fives, four ones
- Ⓑ two tens, one five, nine ones
- Ⓒ one ten, three fives, two ones
- Ⓓ two tens, three fives, four ones

Using Geometry

22. Which of the figures appears to be congruent to the triangle in this droodle?

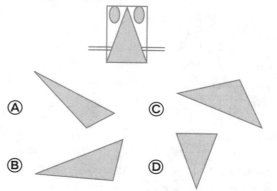

Converting Customary and Metric Measures

20. When Lian's cast first came off, she walked $18\frac{1}{2}$ feet from the living room to the kitchen and then had to rest. How far in inches did she walk?

> 1 foot = 12 inches

- Ⓐ 222 inches
- Ⓑ 216 inches
- Ⓒ 180 inches
- Ⓓ 202 inches

Determining Probability and Averages

23. Lian, Kameko, and their parents drew droodles on index cards and dropped the cards into a large bowl. Look at the chart in the story. If Lian closes her eyes and pulls out a card, whose droodle is she most likely to pull out?

- Ⓐ Lian's
- Ⓑ Kameko's
- Ⓒ Lian's mom's
- Ⓓ Kameko's dad's

Using Algebra

21. Kameko drew this pattern on the front of the droodle book. What figure would complete the pattern?

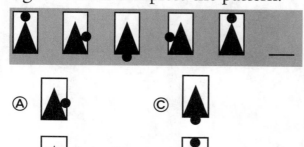

Interpreting Graphs and Charts

24. Lian's leg cast was removed on the first Tuesday in March. She went to ballet practice four weeks after this date. On what date did Lian go to her ballet practice?

- Ⓐ March 31
- Ⓑ April 4
- Ⓒ March 7
- Ⓓ April 5

Strategies 1–12 Final Review 145

PART THREE: Read a Review

Otis wrote a circus review. Read the review. Then do Numbers 25 through 36.

Last night, my family and I took the train to the city to see the Big Apple Circus. This circus is not like any I have ever seen before. For one thing, we were so close to the performers that we could almost reach out and touch them. No one in the audience sits more than 50 feet away from the ring.

Every year, the Big Apple Circus is different. This year, it stars Bello Nock. Bello is a very funny clown with bright red hair that sticks way up in the air. Bello has been a member of the circus since he was 5. His family has been a part of the circus for a long, long time.

In one act, Bello jumped from a high board into a pool. Actually, the "pool" is a trampoline, so when Bello hit the "water" instead of splashing, he bounced. Then Bello picked some other people and me from the audience to march in a parade. Of the 26 people he picked, 3 children were too shy to join in. I felt kind of silly, but it was fun to be in the circus for a little while.

There are other funny clowns, too. One named Francesco juggles giant beach balls. His wild-looking clown suit is a xylophone that really works!

The Big Apple Circus has more than clowns. Trapeze artists do triple somersaults. Other people balance on seesaws or stand on their heads on bicycle seats. The circus also has amazing jugglers, marvelous magicians, and wonderful musicians.

It has elephants named Anna May, Amy, and Ned, who prance around like ballerinas. And it has Violetta's dogs, who race around and jump through hoops. I found out that the Big Apple Circus treats all its performing animals with care and gentleness. Trainers make sure that their animal partners are well fed, rested, and healthy.

If you want an afternoon or an evening of laughter and thrills, see the Big Apple Circus. It will be in town until June 4. Tickets range from $12.00 to $20.00 for afternoon shows, and from $13.00 to $32.00 for evening shows.

Building Number Sense

25. Otis counted 142 people on the train to the city, but he forgot to count himself. How many people were on the train?

- Ⓐ 150 people
- Ⓑ 141 people
- Ⓒ 143 people
- Ⓓ 140 people

Using Estimation

26. Otis learned that during its two-week stay, the circus sold 34,893 tickets. Which of these is 34,893 to the nearest thousand?

- Ⓐ 34,900
- Ⓑ 34,890
- Ⓒ 4,000
- Ⓓ 35,000

Applying Addition

27. At the circus, Otis bought a poster for $5.29, a circus flag for $3.00, and a postcard of one of the circus elephants for 94¢. How much did Otis spend at the circus?

- Ⓐ $10.14
- Ⓑ $9.23
- Ⓒ $8.05
- Ⓓ $10.23

Applying Subtraction

28. The most expensive afternoon ticket costs $20.00. The most expensive evening ticket costs $32.00. How much more is the most expensive evening ticket?

- Ⓐ $10.80
- Ⓑ $12.00
- Ⓒ $18.00
- Ⓓ $8.00

Applying Multiplication

29. The circus will travel to different cities in June and July. Each of those months, they will perform for 4 weeks, putting on 9 performances a week. What is the total number of performances for June and July?

- Ⓐ 72 performances
- Ⓑ 36 performances
- Ⓒ 63 performances
- Ⓓ 45 performances

Applying Division

30. Otis's seat was in Section A, Row 2. Section A has 60 seats, and there are 10 seats in each row. How many rows are there in Section A?

- Ⓐ 3 rows
- Ⓑ 6 rows
- Ⓒ 5 rows
- Ⓓ 8 rows

Converting Time and Money

31. Children who are part of the circus go to school from the time shown on the first clock to the time on the second clock. How long do the children go to school?

    Ⓐ 5 hours, 20 minutes
    Ⓑ 6 hours
    Ⓒ 6 hours, 30 minutes
    Ⓓ 7 hours

Using Geometry

34. During the parade, Otis followed Bello all around the cubes that the elephants would use for their tricks. Look at the drawing that Otis made. How many faces did he see on the cubes when he marched around them?

    Ⓐ 15 faces
    Ⓑ 18 faces
    Ⓒ 12 faces
    Ⓓ 9 faces

Converting Customary and Metric Measures

32. Otis sat 9 meters from the ring. What was his distance from the ring in centimeters?

    Ⓐ 990 centimeters
    Ⓑ 90 centimeters
    Ⓒ 190 centimeters
    Ⓓ 900 centimeters

Determining Probability and Averages

35. If a juggler closes her eyes and picks 2 balls from a box, what two colors is she equally likely to pick?

Color	Number	Color	Number
Purple	9	Green	11
Pink	8	Blue	9

    Ⓐ purple, blue Ⓒ green, blue
    Ⓑ purple, pink Ⓓ pink, green

Using Algebra

33. Which of these number sentences would you use to find how many people marched with Bello?

    Ⓐ 26 + □ = 29
    Ⓑ 26 + 3 = □
    Ⓒ 26 − 3 = □
    Ⓓ □ − 3 = 26

Interpreting Graphs and Charts

36. Which of Otis's circus souvenirs is over 3 and up 4?

    Ⓐ program
    Ⓑ ticket
    Ⓒ flag
    Ⓓ balloon

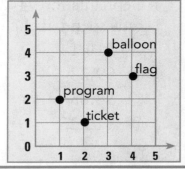

PART FOUR: Read an Article

Read the article about colors found in nature.
Then do Numbers 37 through 48.

Plants and animals are often very colorful. They have color for a reason. Some plants have bright colors to attract animals. Think about the rose. The rose is a beautiful flower with soft, colored petals. This flower attracts the bee that flies near looking for food. The bee obtains nectar and pollen as it touches the flower. But what does the bee do for the flower? This tiny insect carries the flower's pollen away to make new plants. The rose and the bee help each other. They are important to each other for staying alive. Think about the colors of other flowers such as the daisy, the sunflower, and the lily. They, too, attract insects and other animals that search for food. As a result, these creatures also carry pollen or seeds to other places.

Some animals have bright colors to warn other animals to stay away. Think about the ladybug. This tiny creature has a bright color. Its color warns other animals that ladybugs have an awful taste. The bright color helps to keep the ladybug safe. Some animals have bright colors to attract a mate. This is true of several kinds of male butterflies and birds.

Some animals need to hide to be safe. These animals often have dull colors that blend easily with their surroundings. Think about the grasshopper. Its color is brown or green. The color enables the grasshopper to hide in grasses. The zebra has stripes. A leopard has spots. Both animals are able to hide in the shadows of the forest where they live. Stripes and spots break up the outline of an animal's body. These marks and the dull colors keep the animals hidden and safe.

Strategies 1–12 Final Review

Building Number Sense

37. The number of honeybees that visited a rose on Tuesday is the same as the odd number that appears in the box. What is the odd number?

> 16, 31, 24, 18

- Ⓐ 16
- Ⓑ 31
- Ⓒ 24
- Ⓓ 18

Using Estimation

38. To the nearest ten, Clive estimated the number of sunflower seeds in a bag to be about 200. What could be the actual number?

- Ⓐ 215
- Ⓑ 182
- Ⓒ 204
- Ⓓ 173

Applying Addition

39. Marylee has two gardens. Today, in the flower garden, she counted 11 ladybugs. Then, in the vegetable garden, she counted 16 ladybugs. What is the total number of ladybugs that she counted?

- Ⓐ 27 ladybugs
- Ⓑ 21 ladybugs
- Ⓒ 29 ladybugs
- Ⓓ 17 ladybugs

Applying Subtraction

40. A herd of 18 zebras roamed the grassland one afternoon. As evening approached, 7 zebras wandered away. How many zebras remained?

- Ⓐ 6 zebras
- Ⓑ 11 zebras
- Ⓒ 25 zebras
- Ⓓ 13 zebras

Applying Multiplication

41. Each bee has 6 legs that can carry pollen. What is the total number of legs found on 12 bees?

- Ⓐ 18 legs
- Ⓑ 24 legs
- Ⓒ 48 legs
- Ⓓ 72 legs

Applying Division

42. Marylee planted daisy plants in groups of 3 plants. If Marylee planted a total of 39 daisy plants, how many groups did she plant?

- Ⓐ 15
- Ⓑ 9
- Ⓒ 16
- Ⓓ 13

Converting Time and Money

43. A single rose cost $2.30. What coins did Clive use to pay the exact cost for a rose to give to his grandmother?

- Ⓐ 6 quarters, 7 dimes, 4 nickels
- Ⓑ 5 quarters, 9 dimes, 6 nickels
- Ⓒ 8 quarters, 2 dimes, 3 nickels
- Ⓓ 7 quarters, 5 dimes, 1 nickel

Using Geometry

46. Marylee's garden is shaped like a pentagon. Which of these figures is a pentagon?

- Ⓐ
- Ⓑ
- Ⓒ
- Ⓓ

Converting Customary and Metric Measures

44. A grasshopper jumped a distance of 10 feet. How many inches did the grasshopper jump?

- Ⓐ 160 inches
- Ⓑ 120 inches
- Ⓒ 140 inches
- Ⓓ 100 inches

Determining Probability and Averages

47. On Tuesday Clive saw 6 butterflies, on Wednesday he saw 8 butterflies, and on Thursday he saw 4 butterflies. What is the average number of butterflies Clive saw each day?

- Ⓐ 5 butterflies
- Ⓑ 7 butterflies
- Ⓒ 4 butterflies
- Ⓓ 6 butterflies

Using Algebra

45. The missing number in the number sentence is the same as the number of colors in a rainbow. What is the number?

$$13 + y = 20$$

- Ⓐ 7
- Ⓑ 9
- Ⓒ 6
- Ⓓ 8

Interpreting Graphs and Charts

48. Marylee's calendar has 31 days. The first day is on a Monday. What day of the week is the last day?

- Ⓐ Saturday
- Ⓑ Sunday
- Ⓒ Friday
- Ⓓ Wednesday

Strategies 1–12 Final Review

Curriculum Associates 4-Step Mathematics Program

Purpose	Series (Books A–H)	Objectives
Diagnose	*Comprehensive Assessment of Mathematics Strategies (CAMS® Series)*	• to identify a student's level of mastery for each of 12 math strategies (8 strategies in Book A) • to develop effective practices with self-assessment and goal-setting
Teach	*Strategies to Achieve Mathematics Success (STAMS® Series)*	• to provide targeted strategy-specific instruction and practice to students learning key math concepts • to broaden student proficiency in error analysis
Extend	*Extensions in Mathematics™ Series*	• to strengthen students' problem-solving skills and math writing skills using graphic organizers • to expand on the 12 standards-based strategies (8 strategies in Book A) promoted in the *CAMS® Series* and *STAMS® Series*
Assess	*Comprehensive Assessment of Mathematics Strategies II (CAMS® Series II)*	• to assess students' math proficiency at the conclusion of the instructional period • to continue the development of effective practices with self-assessment